Shape in Chemistry

Shape in Chemistry

An Introduction to Molecular Shape and Topology

Paul G. Mezey

University of Saskatchewan

VCH

Paul G. Mezey
Mathematical Chemistry Research Unit
Department of Chemistry and Department of Mathematics
University of Saskatchewan
Saskatoon   S7N 0W0
Canada

This book is printed on acid-free paper. ∞

Library of Congress Cataloging-in-Publication Data
Mezey, Paul G.
    Shape in Chemistry : an introduction to molecular shape and
topology / Paul G. Mezey.
        p.   cm.
    Includes index.
    ISBN 0-89573-727-2
    1. Molecular structure.   2. Topology.   I. Title.
QD461.M583   1993
541.2'2 – dc20   93-15622
                                                     CIP

Printed in the United States of America

ISBN 0-89573-727-2 VCH Publishers, Inc.
ISBN 3-527-27932-6 VCH Verlagsgesellschaft

Printing History
10 9 8 7 6 5 4 3 2

Published jointly by

VCH Publishers, Inc.   VCH Verlagsgesellschaft mbH   VCH Publishers (UK) Ltd.
220 East 23rd Street   P.O. Box 10 11 61   8 Wellington Court
New York, NY 10010-4606   69451 Weinheim   Cambridge CB1 1HZ
                            Germany   United Kingdom

Preface

Shape is one of the most fundamental concepts in natural sciences. The recognition of shape is one of the dominant aspects of human visual perception and it plays a primary role in the process of understanding natural phenomena. Yet, in our everyday experiences this very concept is somewhat subjective and vaguely defined; to each observer a different aspect of form may appear important. To some degree, shape is in the eye of the beholder. However, visual inspection is not the only method for characterizing shapes: it is possible to describe shapes precisely and unambiguously by geometrical and topological methods. Such precise formulations lead to rigorous mathematical methods for shape characterization and to algorithmic, computer based approaches, thereby eliminating the subjective element of shape analysis. The spectacular developments of computer hardware and software along with the revolution in computer graphics techniques have made their mark on the evolution of shape analysis and in making the traditionally rather elusive concept of shape more accessible to rigorous scientific inquiry. Today, shape is in the computer of the beholder.

Shape in chemistry appears on many levels. The shapes of crystals, the shapes of titration curves, the shapes of spectral lines, the shapes of potential energy functions or the multidimensional shapes of potential energy hypersurfaces are some examples. However, few chemists would dispute that the most important shape problem of chemistry is that of the three-dimensional shapes of molecules. The study of molecular shape and molecular shape changes is fundamental to our understanding of chemical properties and reactions.

The purpose of this book is to acquaint the reader with the topological methods of the description and analysis of shapes, in particular, the three-dimensional shapes of molecules. The topological approach is generally applicable to all aspects of shape in chemistry, hence the title of the book appears justified, although our focus will be on the central shape problem in chemistry: on molecular shape.

Throughout the book, the emphasis is placed on the topological characterization of the shapes of the "fuzzy", three-dimensional *bodies* of electronic densities of molecules, as opposed to the more conventional stereochemical description of the shapes of molecular *skeletons* obtained when representing molecular structures by formal chemical bonds. Topological shape analysis methods are of relevance in both theoretical studies of molecular shapes and in a variety of applications. The mathematical treatment is rather elementary and self-contained. In particular, the reader needs no special background in topology, and the usual undergraduate exposure of a chemistry student to mathematics is all that is required to follow the book. The necessary topological concepts are introduced gradually, starting at the familiar level of ordinary stereochemistry, leading through simple, pictorial concepts to the more advanced topological shape analysis methods. The topological shape group and shape code methods provide a precise shape description of fuzzy electronic charge density distributions, numerical measures of molecular shape similarity, and shape complementarity, as well as the computational means for a rigorous shape classification of molecules. These methods are also applicable in the study of shapes of electrostatic potentials, Van der Waals surfaces, solvent accessible surfaces, and formal molecular surfaces and molecular bodies defined by some other criteria. These techniques provide a basis for nonvisual computer analysis of both *static* and *dynamic* shape similarity and shape complementarity in sequences of molecules. Such analyses are important in drug design and Quantitative Structure Activity Relations (QSAR) studies, leading to a shift in emphasis from the stereochemical *bond structure* to the three-dimensional *shape of molecular bodies* and to Quantitative Shape Activity Relations (QShAR).

Motivation for developing rigorous, nonsubjective molecular shape analysis methods comes from nearly all branches of chemistry. Most physical and chemical properties of molecules, reactivity, reaction mechanisms, biochemical activity, and drug action are strongly dependent on molecular shape. Shape is a fundamental property of both macroscopic and microscopic objects. However, in the case of molecules, where quantum mechanics is necessary for a valid description of electron distributions, the concept of shape is somewhat different from that of macroscopic objects. Due to the Heisenberg uncertainty relation, there is an inherent fuzziness associated with molecular electron distributions, which renders the usual, geometrical concepts of macroscopic shape characterization less than ideal for molecular problems. A generalization of geometry is topology, which is a more appropriate tool for describing quantum mechanical objects such as molecules. Topology is also an efficient tool for extracting the essential information from complicated objects, such as the large scale shape features of tertiary structures of proteins. The power of topology to focus on the essential features of complex structures is exceptionally useful when analyzing molecular shape. For both quantum mechanical and practical, computational considerations, topology appears to be an ideal tool for molecular shape analysis.

Research efforts to develop rigorous molecular shape analysis methods based on three-dimensional topology (which should never be confused with graph theory) have a relatively short history. Nevertheless, the concept of molecular shape is nearly as old as the concept of molecule. The importance of stereochemistry and conformation analysis has led to an early recognition of the need for models of three-dimensional molecular shapes, and there have been many studies on the subject. Most of these studies have been based on models of bond skeletons or space filling models used as representatives of molecular shapes. This book, however, has a different perspective and a different aim. Our goal is to introduce the reader to the topological concepts and methods of precise shape characterization that, with the use of computers, are applicable for direct, nonvisual description and analysis of general molecular shapes. Systematic and rigorous analysis and characterization of shapes of molecules are particularly important in computer-aided drug design and molecular engineering, where the level of detail required often varies, depending on the actual molecular problem. Consequently, special attention will be given to topological methods suitable for the study of both the highly detailed and the crude, large scale shape features of molecular systems, in particular, biologically important molecules and macromolecules such as peptides and proteins. Whether applied to small or large molecules, most of the techniques described here are based on some of the elements of topology, and the book provides a simple, pictorial introduction to all the topological tools necessary for the subjects discussed. Since the book is aimed at a wide audience, including advanced undergraduates, graduate students, nonspecialist organic, physical and medicinal chemists, and research workers in various aspects of QSAR, QShAR and pharmacological drug design, the mathematical description is kept at a simple and easily comprehensible level. It is my hope that the reader will acquire the (presumably addictive) habit of thinking about molecules and reactions in topological terms, which may lead to further advances in exploiting the seemingly unlimited potential of topological concepts, the vast collection of the available topological results, and the emerging topological computational techniques in chemistry.

The structure of this book reflects the author's desire to bring the reader to an early appreciation of the power of topology in chemistry. In Chapter 1, the concept of molecular shape is discussed in terms of the conventional, intuitive ideas of stereochemistry. Internuclear distances, symmetry and chirality give only a partial characterization of molecular shape, and the fact that molecules are nonrigid, dynamic entities provides the motivation to search for more general approaches to shape description. In Chapter 2, the quantum chemical concept of molecular shape is reviewed briefly. Here the Heisenberg uncertainty relation plays a prominent role by providing the theoretical basis for the topological approach. A molecular body can be represented by a series of molecular isodensity contours (MIDCO's) for the entire range of chemically relevant electron density values. A systematic treatment of the quantum chemical information leads to a novel topological description of bonding in molecules: the Density Domain Approach (DDA) to chemical bonding. The topological density domains are the basis for a quantum chemical definition and generalization of the functional groups of chemistry.

Chapter 3 is entitled: "Applied Topology: The Mathematics of the Essential". This is a brief, mostly pictorial review of the most basic concepts of topology relevant to the problem of molecular shape, relying on many figures and illustrations and avoiding most of the formal mathematics. The precise, mathematical formulation of the molecular shape analysis problem is discussed in detail in the original references, listed at the end of the book.

In Chapter 4, those physical properties and molecular models are discussed which may serve as the representatives of a formal molecular body and formal molecular surfaces, including shape representations of macromolecules and protein folding. These properties and models include charge density, electrostatic potential, Van der Waals (fused spheres) models, solvent accessible surfaces, as well as ribbon and polyhedral representations of the large scale shape features of biopolymers. The associated contour surfaces serve as the basis for the introduction of the topological shape groups, shape codes, shape graphs, and shape matrices, as well as the shape globe invariance maps for quantitative molecular shape comparisons in Chapter 5. Also in Chapter 5, topological approaches are described for the computer representation of folding patterns and chirality properties of chain molecules.

In Chapter 6, the merits of visual, computer graphics methods and nonvisual, algorithmic shape analysis methods are compared, and the principle of Geometrical Similarity as Topological Equivalence (GSTE) as well as the Resolution Based Similarity Measures (RBSM) are reviewed. Topological approaches to the quantification of molecular similarity and complementarity are presented, formulated as molecular shape similarity measures and shape complementarity measures, respectively. In Chapter 7, special methods and computer algorithms are described for molecular similarity analysis, designed for applications in Quantitative Shape-Activity Relations (QShAR) studies. Computational methods are proposed for main effect and multiple side effect analysis by shape correlations, with applications in drug design, and molecular engineering. In the final section, Chapter 8, the concepts of approximate symmetry and various generalizations of symmetry are discussed, including symmetry deficiency measures, syntopy and symmorphy.

I am most grateful for the help I have received from many colleagues, students, and friends in writing this book. In particular, I thank the contributions of former and current members of the Mathematical Chemistry Research Unit of the University of Saskatchewan, Drs. S. Arimoto, G.A. Arteca, I. Bálint, J.Y. Choi, G. Heal, V. Jammal, X. Luo, J. Pipek, C. Soteros, K. Taylor, and P.D. Walker. Special thanks are due for H. Acton, G. Gates, R. Gerwing, M. Kudel, M. Mawer, and Dr. P.D. Walker for careful reading of various versions of the manuscript. The financial help received from the Natural Sciences and Engineering Research Council of Canada (NSERC) in the form of operating and strategic grants for the development of some of the shape analysis methods reported, and the support received from The Upjohn Company are gratefully acknowledged.

Paul G. Mezey
Saskatoon
June 1993

Contents

INTRODUCTION

How would the world appear to you if you were of the size of a water molecule? Let us take an imaginary journey in that world. Imagine that you are a molecule and you are living among other molecules. Think of these molecules as intelligent beings; they are able to detect each other, to describe each other, they understand mathematics and quantum mechanics, and in general, they know well the physics and chemistry governing their behavior. Try to imagine what concepts and methods these intelligent molecules would use to describe each other, to characterize their shapes, their conformational changes, and their reactions. As macroscopic human beings, we might find it difficult to understand the mindset of our microscopic molecular friends. But there are some clues, and in this book an attempt will be made to give an introduction to a subject that intelligent molecules themselves could find important: the topological characterization of molecular shape.

Back to our real world, for an understanding of molecular properties we must use our concepts and the accumulated information obtained by macroscopic means. Throughout the development of chemistry, pictorial models of molecules have guided chemical intuition and have often led to important discoveries. The ways in which chemists imagine molecules have a major impact on our understanding of molecular behavior and chemical reactions. It is not surprising that macroscopic analogies are often used, and one is tempted to imagine molecules as microscopic "bodies", with well defined size and shape properties not fundamentally different from those of objects we encounter in our everyday life. However, we must be cautious when using such analogies. Our everyday experiences with macroscopic bodies are well described by classical mechanics; therefore, macroscopic analogies are essentially classical mechanical. By contrast, molecules cannot be described by classical mechanics alone. Quantum mechanics plays a much more important role on the microscopic, molecular level, than on the macroscopic level of ordinary objects we can easily perceive with our senses. During the course of human evolution there has been no apparent need for direct observation of microscopic, quantum mechanical phenomena; accordingly, human senses, hence also human imagination, are poorly prepared to deal with some of the quantum mechanical properties of molecules. We are prone to think and reason using macroscopic examples which are, by their nature, classical mechanical and in several aspects alien to the microscopic world of molecules. Consequently, in chemistry, macroscopic analogies may be very misleading. Although some analogies may work well, perhaps even too well in many instances, it is much too easy to attempt to stretch the application of such analogies beyond their range of validity. If a macroscopic analogy is proven to be useful in the interpretation of some results, then it is tempting to disregard the fact that it is only an analogy and to use it as if it were an exact representation of reality. In particular,

1

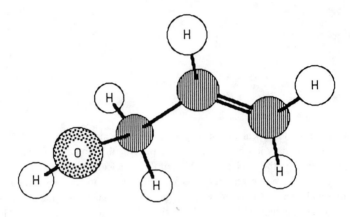

Figure I.1 A ball and stick model of the allyl alcohol molecule is shown. The formal bonding pattern is well recognizable in such "skeletal" models, however, the actual three-dimensional shape of the fuzzy "body" of the molecular electron density, ultimately responsible for chemical bonding, is not well represented.

the concept of molecular shape is not a trivial one; it cannot be represented exactly by classical, macroscopic models, and it must reflect the microscopic, quantum mechanical nature of molecules.

Most of chemistry is taught using chemical formulas and structural diagrams expressed in terms of chemical bonds. These bonds are usually drawn as lines, and molecular shapes are often represented by stereochemical, structural formulas that are line drawings, with indications of the three-dimensional directions of chemical bonds. From the early days of stereochemistry [1-3], such stereochemical bond diagrams have proved extremely successful in explaining isomerism and the structural properties of a wide variety of molecules [4-16], molecular optical activity and chirality relations [17-58,59-72], as well as more general symmetry properties of molecules [73-82]. Important advances have been made in the study of stereochemical changes and have lead to systematic approaches to conformation analysis [83]. Simple stereochemical bond diagrams and "ball and stick" models, however, do not convey the actual space requirements of molecules and functional groups. By contrast, various space filling models [84] of partially fused spheres, Van der Waals surfaces [85-88], or electron density contour surfaces [89-92] provide a fundamentally different and more realistic picture of molecular shapes.

The introduction of Valence Bond theory has motivated the search for structural regularities that can be interpreted by models of local electronic features, such as the powerful model of Valence Shell Electron Pair Repulsion [93,94] theory. Alternative approaches, based on Molecular Orbital theory, have led to the discovery of important rules, such as the Woodward-Hoffmann orbital symmetry rules [95] and the frontier orbital approach of Fukui [96,97]. As a result of these advances and the spectacular successes of *ab initio* computations on molecular

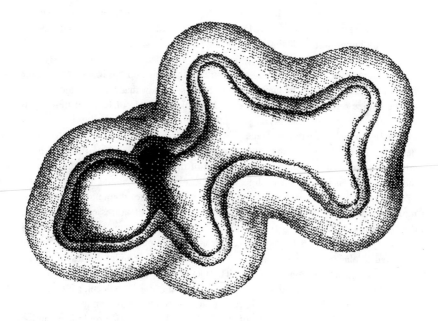

Figure I.2 The three-dimensional, fuzzy "body" of the charge density distribution of allyl alcohol can be represented by a series of "nested" molecular isodensity contours (MIDCO's). Along each MIDCO the electronic density is a constant value. Three such MIDCO's are shown for the constant electron density values of 0.2, 0.1, and 0.01 (in atomic units), respectively. A contour surface of lower density encloses surfaces of higher density. These MIDCO's are analogous to a series of Russian wooden dolls, each larger doll enclosing a smaller one. These *ab initio* MIDCO's have been calculated for the minimum energy conformation of allyl alcohol using a 6-31G* basis set.

structures, the concept of molecular shape has become a truly three-dimensional, although still primarily *geometrical* notion. This has marked an important stage in the evolution of the concept of molecular shape, that can be attributed in part to a more universal appreciation of the Molecular Orbital approach [98], as well as to a need for a better understanding of biochemical processes [99].

An example of a skeletal ball and stick model of the allyl alcohol molecule is shown in Figure I.1. Whereas the formal bonding pattern is well recognizable in the Figure, the actual shape of the molecular electron density distribution ultimately responsible for bonding is very different. In Figure I.2, the three-dimensional charge density distribution of allyl alcohol is represented by a series of molecular isodensity contours (MIDCO's) that are formal surfaces along which the electron density is constant. Three such "nested" MIDCO's are shown in Figure I.2, for the arbitrarily chosen constant electron density values of 0.2, 0.1, and 0.01 a.u. (atomic units), respectively, where the contour surfaces of lower density enclose those of higher density. The shape of the formal molecular body can be represented by an infinite family of such nested MIDCO's, for all possible electron density values. A series of nested MIDCO's is analogous to a series of Russian wooden dolls, each larger doll enclosing a smaller one.

The fact that geometrical approaches are not ideally suited to represent the fuzzy nature of quantum chemical electron distributions and the inherent, quantum mechanical uncertainties of nuclear arrangements has not made yet a major impact on the choice of the most commonly used chemical models. Chemical thinking is still dominated by the traditional structural formulas of bond *skeletons*, and the three-dimensional, formal molecular *bodies* receive less attention than they deserve. In most molecular beauty contests the jury of chemists still pays more attention to the X-ray picture of the skeleton of the contestants than to the shape of their three-dimensional bodies. Molecules themselves ignore human bias and appear to have more refined taste: molecular recognition and the interactions of reacting molecules are dominated by the three-dimensional shape features of the bodies of their electronic charge density distributions. In the conventional interpretation of chemistry, the shapes of these fuzzy electron distributions are still much too often relegated to play a role that appears only secondary to the simpler but less revealing skeletons of structural bond formulas. The fact that the peripheral regions of fuzzy bodies of electronic charge densities have a dominant role in molecular interactions is well understood, but it has not yet fully transformed chemical thinking.

The three-dimensional shape of this fuzzy body of the electronic distribution has many important features not revealed by the simple, skeletal ball and stick model. One of the most important tasks of topological shape analysis of molecules is the precise analysis and concise description of the three-dimensional electronic charge distributions, such as that illustrated by the selected MIDCO's of allyl alcohol in Figure I.2. Various methods and computational techniques of such topological shape analyses are discussed in detail in this book.

The evolution of the concept of molecular shape has been mirrored by a parallel evolution of chemical models and representations in chemistry [100], ranging from simple, classical analogies through advanced models of stereochemistry, conformational analysis and MO theoretical descriptions of molecules to the most recent approaches [101-108]. Early mathematical models used for the description of molecules and chemical reactions all had their roots in classical-mechanical, geometrical concepts. Early stereochemistry was formulated in terms of three dimensional (3D) structural formulas, where chemical bonds of precise directions in 3D space have accounted for molecular structures. Chemical reactions have been perceived as processes in which such bonds are created and destroyed. Again, geometrical concepts, predominantly lines with well defined orientations, have formed the basis of modeling chemical reactions. Such models have been of great utility, although they have also led to an unnecessarily oversimplified perception of chemical bonding. The spectacular development of quantum chemistry and the major advances in the computation of both *ab initio* and semi-empirical MO wavefunctions have initiated a critical re-evaluation of some of the fundamental concepts of chemistry, such as chemical bonding, stereochemical properties of molecules, and molecular shape. All these properties are determined by fuzzy, quantum mechanical electron distributions, where the classical, geometrical concept of a precise position of an electron within a molecule has become meaningless. To a lesser degree, nuclear positions are also subject to quantum mechanical fuzziness within a molecule, and this has important consequences for stereochemistry. In order to describe this inherent fuzziness, it is natural to relax the classical, geometrical models and replace them with models which can account for the nongeometrical nature of electronic and nuclear positions within molecules. One such model, based on a quantum mechanically motivated topological approach, has been proposed within the framework of the Chemical Topology Program (CTP), or in short, the Topology Program (TP) [103-112]. This program is a general framework for a family of concerted research projects suggested for a topological description of molecules and chemical reactions. The two main components of the Topology Program are Molecular Topology, MT, [103,106-112], and Reaction Topology, RT, [104-107,111].

Topology is a branch of modern mathematics [113-122] that appears ideally suited for a detailed description of molecules, conformational changes, and chemical reactions. Topology deals with the continuity of functions, the connectedness of objects, and with *shape,* both in a concrete, three-dimensional setting and in a more abstract sense. Topologists are often teased for being unable to distinguish the doughnut from the coffee-cup, since both objects have just one hole, and if both objects are made of soft, malleable material, then each can be deformed continuously and reversibly into the other, without tearing or gluing any of their parts. Throughout these reversible deformations, continuity is preserved and, consequently, the two objects are topologically equivalent. Topological techniques are suitable for describing aspects of shape much more detailed than the mere presence of holes, in fact, any finest detail of molecular shape can be analysed within a rigorous topological framework.

The development of methods and computational techniques of the Topology Program described in this book has been based on the physical properties of molecules. This has required the introduction of new topological approaches for shape analysis. Those readers who are mathematically inclined may note that these new approaches are different from the methods used for the more general, abstract shape problem of topology [123-154], where the emphasis is placed on general techniques for the characterization of objects which may have pathological local behavior. In particular, the algebraic groups of the molecular Shape Group Method (SGM) have been specifically designed for the class of (nonpathological) shape problems of molecular quantum mechanics [155-158], and should not be confused with the abstract shape groups [130,143] developed from Borsuk's topological shape theory [123-127].

Topology is one of the most loosely used words in the contemporary chemical literature and is often confused with graph theory [159]. Whereas topology typically deals with continuum problems, graph theory typically deals with discrete problems. Topology describes continuous shapes of objects such as three-dimensional bodies and their changes, whereas graph theory is a powerful tool for representing binary, that is, yes - no type relations, among various entities. Of course, many of the discrete results of a three-dimensional topological shape analysis can be represented by graph theoretical means.

Topology provides unifying links among the most diverse branches of mathematics, but it is only in the recent past that some of the elementary results of differential topology and algebraic topology have found applications within chemistry. In view of the exceptionally strong parallels between the available repertoire of topological methods and the physical nature of molecules and reactions, it is somewhat surprising that the discovery of the true range of powers of topology in chemical applications has occurred only recently. In the context of this book, topology provides a particularly suitable mathematical approach to molecular shape problems within both quantum chemistry and molecular modeling. In fact, topology provides a link between these two sub-disciplines of modern chemistry. These connections and the general applicability of topology in chemical shape analysis are expected to provide many new insights as well as new challenges.

Although the nongeometrical nature of quantum chemistry is widely recognized, the classical, geometrical viewpoint still seems to overshadow somewhat the information gained from quantum chemistry. The emphasis is often placed on those aspects of quantum chemistry which can be interpreted easily in terms of conventional, geometrical concepts, such as lines of chemical bonds between atom pairs. The dominant means of communicating chemical ideas has remained the concept of chemical bonds thought of as lines, even after various diffraction experiments and computational quantum chemistry have clearly established the facts that

1. molecular electron distributions are rather diffuse and somewhat fuzzy entities, and
2. only a small fraction of the full 3D information can be represented by oriented, geometrical lines.

In the recent past, a considerable research effort has been focused on searching for quantum chemical models that are expected to reconcile quantum mechanics with most of the traditional, classical and geometrical concepts of chemistry. These efforts have been somewhat at the expense of the study of the global, delocalized shape aspects of 3D bodies of actual electron distributions.

Of course, the stereochemical importance of 3D models of formal molecular bodies of electron distributions was recognized rather early, but for long these electron distributions have been viewed more or less as the dressing on a 3D structural skeleton of bonds, and the mathematical tools (precisely defined geometrical surfaces and bodies) have remained classical and geometrical. Mechanically constructed 3D "space filling" molecular models, capable of formal single bond rotations, represented an improvement, but they too could provide only a fraction of the full, 3D information on real molecules in an easily comprehensible manner. Perhaps a more significant limitation was the fact that even the most versatile of these mechanical models could be modified and readjusted only in very restricted ways. For example, most mechanical models could not account for bond stretching while preserving the continuity of a formal molecular surface. Modeling chemical reactions, a problem considerably more complicated than modeling chemical species, has faced similar difficulties on a larger scale.

The advances in computer hardware and in the methods of molecular graphics have led to a much better appreciation and to a variety of applications of the new knowledge, concepts, and insights gained from quantum chemistry. One of the important developments is a true breakthrough in the computer modeling of molecules and chemical reactions. It is now a routine task to generate, modify, and manipulate virtually without limitation the computer displays of 3D molecular models. All these can be done with an ease and speed that allows a quick comprehension and an effortless understanding of the wealth of detail of shapes of real, 3D molecules. Computer-based molecular modeling has revolutionized the treatment and analysis of 3D molecular information. This resulted in a new appreciation and a better understanding of stereochemistry, both on a conceptual level and on the level of applications to important molecular problems. Instead of thinking in terms of one-dimensional entities, such as lines representing chemical bonds, the modern approaches to interpreting molecular properties increasingly rely on computer models of 2D molecular contour surfaces and 3D molecular bodies. This development may be more natural than it seems for reasons that go far beyond chemistry. Before the evolution of the capacity for abstract thinking needed for two-dimensional representations in planar drawings and writing, prehistoric human perception and imagination were probably dominated by three-dimensional shape properties. For a long time following the introduction of writing, the primary aid in learning and understanding became the two-dimensional, printed paper. This has led to a severe bias in favor of 2D representations and thought processes involving 2D images. The human skills for 2D thinking, still dominating our education, have developed in part at the expense of practicing 3D imagination. But, with the advent of computer modeling, our innate capacity for three-dimensional thinking and understanding received a well deserved boost. Indeed, by shifting the emphasis from 1D bonds to 2D contour surfaces and 3D molecular bodies, a "dimension revolution" has occurred in chemical understanding.

The new computer capabilities of treating 3D molecular models require a new set of mathematical tools for their analysis. Geometry, in particular differential geometry, is the natural tool for the detailed description of precisely defined, smooth surfaces and bodies. This tool is very powerful as long as the molecular model used is defined in terms of classical mechanical analogies, for example, by assuming the existence of a molecular surface, and an associated molecular body with a well defined boundary. Van der Waals surfaces (VDWS's) are models based on the above assumption and they are exceptionally useful tools for the approximate representation of molecules, particularly macromolecules. However, real molecules obey the laws of quantum mechanics and are subject to the Heisenberg uncertainty relation. Consequently, in the strict sense, one has no unique, precise, physical definition for the nuclear configuration of a molecule, neither for molecular surfaces, nor for molecular bodies. No molecule exists, not even for an instant, with any precise nuclear geometry. Hence, in the strict sense, nuclear configuration is a nonphysical concept. Instead, an infinite family of formal nuclear conformations (that is, an infinite family of geometrical models) within a neighborhood of a potential minimum can be considered as representing a molecule. In a rigorous description, only some of the essential, common features of the whole family of possible geometrical models can be regarded as an expression of physical reality. The common, invariant features within any family of geometrical models are usually topological properties, called topological invariants and hence, it is natural to use topological techniques for their analysis.

If an object has a well-defined geometry, then we shall classify it as a *geometrical object.* If an object does not have a precisely defined geometry, but it preserves some topological features, then we shall classify it as a *topological object.* In this context, it is important to keep in mind that an object can have well-defined topology even if it does not have a well-defined geometry. For example, a macroscopic, cut diamond is a geometrical object to a very good approximation: its geometrical descriptors can change only very little without changing the identity of the diamond, for example, by breaking it into two pieces. By contrast, a cat can change its geometrical parameters to a great extent while staying the same cat and preserving its essential topological features. Consequently, a cat is a topological object. These examples also illustrate the role of the level of resolution. If one views the diamond on a microscopic, atomic level of high resolution, then its geometrical features become blurred due to lattice vibrations, and also due to the Heisenberg relation. On this level of resolution the diamond can also be regarded as a topological object.

Molecular topology, the primary aspect of the topology program, is based on the following, fundamental principle: *molecules are not geometrical but topological objects* [106-108]. Molecular topology provides a general framework for molecular shape analysis that incorporates both the static and the dynamic aspects of molecular shape. In a general process of molecular transformation, such as a conformational rearrangement or a chemical reaction, many shape features of a molecule may change but some topological shape properties do remain invariant. A topological, dynamic shape analysis deals with both the variable and the preserved aspects of shape. Both the static and dynamic aspects of shape can be formulated in terms of topological shape invariants (the topological features which do not change in minor conformational changes) and in terms of the families of nuclear arrangements (configuration space domains) within which these topological shape invariants are preserved. The topological shape invariants and the corresponding families of nuclear arrangements (conformational domains) can be characterized topologically and algebraically, leading to simple algorithms and computer programs for nonvisual, dynamic shape analysis.

Molecular shape analysis has assumed an increasingly important role in biochemical research as well as in pharmaceutical drug design [160-189]. Biochemical applications of computational chemistry, in particular, the large variety of available computer program packages for molecular modeling have added a new dimension to biochemistry. In addition to *in vivo* and *in vitro* experiments, it is now possible to carry out important studies *"in computo",* that is, by computer modeling, an aspect of modern biochemical research that may reduce (but not eliminate) the reliance on animal experiments. Medicinal chemistry, pharmacology, the study of the main and side-effects of potential pharmaceuticals, toxicology, environmental chemistry, pesticide and herbicide research for the development of new agricultural chemicals as well as polymer chemistry, supramolecular chemistry, the research efforts for the development of new industrial materials, are all expected to benefit from these advances.

Molecular topology [155-158,190-199] presents a systematic framework for general shape analysis methods applicable, in principle, to all molecules. The same framework is also the basis for special shape analysis methods designed to exploit the typical features of some special, distinguished molecular families, such as the folding properties of polypeptides, proteins, and other chain biomolecules. Molecular topology and the associated topological shape analysis approaches form the basis of the present book.

CHAPTER

1

THE INTUITIVE CONCEPT OF MOLECULAR SHAPE

The idea, which seems so obvious today, that the stoichiometry of atomic constitution of molecules alone is not sufficient to determine the identity of molecules, and that the three-dimensional shape of atomic arrangements are of decisive importance in chemistry, has not taken hold easily among chemists. The existence of isomerism [200], three-dimensional stereochemistry [1-3], and conformational diversity (for a review, see [83]) were revolutionary ideas when first proposed and were fiercely resisted by many in the contemporary chemistry community. An excellent summary of the early development of stereochemistry can be found in the historical account of O. B. Ramsay [13]. It is of some interest to note that the first major hurdle chemical thinking had to clear in developing a realistic molecular concept was the idea that molecules have three-dimensional shape properties similar to those of macroscopic bodies. Yet, the newer quantum chemical evidence appears to force us to clear that same hurdle again, this time from the opposite direction (and arguably at a somewhat higher, more difficult level) as the restrictions on the validity of macroscopic shape analogies are becoming more evident. The *shape features of molecules are different from those of macroscopic bodies* in many important aspects. Macroscopic analogies served chemists well on the first occasion, but today, excessive reliance upon them appears to hinder and restrict our understanding of the true shape properties of molecules.

1.1 Shape in Stereochemistry

Within the commonly accepted terminology of stereochemistry, molecular shape is understood as being the three-dimensional arrangement of formal chemical bonds. Most approaches toward the visualization of stereochemical arrangements of atoms

10

within a molecule use ball-and-stick models and similar representations, where the emphasis is placed on the bonding pattern, that is, on the way atoms appear linked to one another, eventually forming a molecule. There are good reasons for the success of this approach. In most instances, the individual steps in the synthesis of larger molecules involve the formation or destruction of just one or two formal chemical bonds and highly concerted, simultaneous formation or destruction of several formal bonds is more the exception than the rule. It is understood that a change in the bonding pattern within a local neighborhood of a molecule may change the electron distribution throughout the whole molecule, though perhaps only slightly. Yet, the qualitative aspects of the presence or the lack of a formal chemical bond between a given atom pair some distance away from the local neighborhood are not expected to change. Some heats of formations can be approximated fairly well by assuming additivity of formal bond energies. Local hybridization schemes account reasonably well for some of the local shape features of directed chemical bonds. Furthermore, the simple hybridization model can be much improved by considering the usual space requirements of lone electron pairs and bonding pairs within the Valence Shell Electron Pair Repulsion (VSEPR) method [93,94], while still preserving the essential aspects of the stereochemical model of bonds between atom pairs. Hence, the concept of chemical bonds, that is, a formal *skeletal model for molecular shape,* is an extremely successful one.

However, the need for a better description of the formal molecular body, the need to account for molecular volume effects, the necessity to describe finer details of changes in electron distributions during conformational changes and chemical reactions, and the requirement of a more precise evaluation of molecular similarity are the factors which have motivated chemists to move beyond the stereochemical skeletal shape concept.

Before the computer age, models suitable for representing formal molecular bodies and volume effects had to be constructed mechanically. The early space-filling models of molecules [84] have provided a better perspective on chemistry and, to the surprise of many chemists, a realization of how compact and closely packed most molecules are. These models have enabled chemists to appreciate the importance of steric hindrance and nonbonding interactions. The concepts of ring strain, rotational barrier, neighbor functional group repulsion interactions, and steric requirements, some of which had often been dismissed as merely speculative, have gradually gained acceptance, due, in part, to the introduction and success of space-filling models. Mechanically constructed molecular models had important roles in chemical discoveries, one prominent example is the insightful use of simple models of nucleotide bases, leading to the discovery of the double helix.

The stereochemical shape concept covers a wide range of possible resolutions, from the details of electron density distributions between pairs of nuclei in relatively small molecules to the structural organization of the tertiary structure of proteins [201-203], the architecture of supramolecular assemblies [204-230], the problems of shape selectivity in reactions of large molecules [231-233], and the intriguing shape features of self-replicating chemical systems [234-239]. In the following chapters we shall discuss various topological shape analysis techniques, suitable for the relevant level of resolution.

1.2 Chirality and Molecular Shape

An object is chiral in the ordinary three-dimensional space if, by translations and
rotations, it cannot be superimposed on its mirror image. Achiral objects are
superimposable on their mirror images. By superimposition we mean a perfect
overlap, that is, we require that the superimposed objects are indistinguishable. The
word "chirality", meaning "handedness", is of Greek origin; *cheir,* or *cheiros* is
Greek for "hand". The human hand is an obvious example of a chiral object.
Chirality properties of molecules have implications in a wide variety of chemical
fields; these range from the basic quantum mechanical properties of simple
molecular systems of a few nuclei and electrons to molecular optical activity,
asymmetric synthesis, the folding pattern of proteins, and the topological chirality
properties of certain catenanes and supramolecular structures. Chirality is an
important shape property, in both the geometrical and the topological sense.
Chirality has been the subject of fundamental studies in various branches of
mathematics. In particular, new developments in a branch of topology, called *knot
theory,* as well as in various branches of discrete mathematics, have led to a novel
perspective on the topological aspects of chirality and to some novel applications to
problems of molecular chirality. Some of the mathematical advances have helped in
the interpretation of many new concepts in theoretical chemistry and mathematical
chemistry. The theoretical advances have motivated novel synthetic approaches
leading to new molecules of exceptional structural properties. Some of the new
developments in molecular chirality have been truly fundamental to the theoretical
understanding and to the actual practice of many aspects of chemistry.

   Chirality is one of the shape features of molecules that has been recognized
early as having important chemical consequences [17-58]. Although in the strict
sense, mirror images of chiral molecules have equivalent *intrinsic* shape features and
they differ only in the way they are *embedded* in the three-dimensional space,
nonetheless, we shall regard three-dimensional shape in a generalized sense by
considering both intrinsic shape features and embedding properties. Ordinary
chirality in three dimensions is a manifestation of the lack of certain point symmetry
elements (i.e., the lack of reflection planes and the lack of S_{2k} type symmetry
elements for k>0).

   Molecular chirality is an energy-dependent property. If sufficient energy is
available, then any chiral molecule can be transformed into its mirror image; if by
no other means, then by a complete dissociation into atoms followed by a reassembly
of the mirror image of the original molecule. What makes chirality properties
special in comparison with other symmetry properties is the fact that chirality is
defined by the *lack* of certain symmetry elements. By infinitesimal distortions of a
general, nonplanar molecule, any point symmetry element can be eliminated. Hence,
a change of point symmetry properties of molecules does not in general require
much energy. In contrast, it usually takes rather substantial, noninfinitesimal
distortions - and hence, also substantial energy - to convert a chiral molecule into
an achiral molecule by attaining a symmetry element of a mirror plane or that of a
suitable S_{2k} axis. Consequently, chirality is a more stable molecular property than
the point symmetry group of a given formal nuclear arrangement.

Chirality appears on many levels in chemistry. The simplest of these is exemplified by the case of a carbon atom with four substituents, *where no two substituents are mirror images of each other.* Such carbon atoms are called *chirogenic* [62]. Note that this condition differs from the more commonly applied condition of having *four different substituents* in the usual Cahn-Ingold-Prelog (CIP) chirality classification scheme [19]. If two substituents R and S are chiral mirror images, then the conventional CIP chirality condition is satisfied for the *achiral* methane derivative CHFRS (with Fluorine F and the two chiral groups R and S as substituents) or for a similar local molecular moiety, a disadvantage when applying the mathematical techniques of chirality for disentangling the local contributions to the overall chirality of a molecule [62]. The chirogenicity condition [62] overcomes some of these difficulties.

Chirality also appears on a somewhat higher level, for example, in molecules which contain no such formal centers of chirality but where the molecule as a whole is chiral. An example of this case is the molecule of hexahelicene, containing a spiral arrangement of six fused aromatic rings. Beyond the local chirality of individual amino acids in a protein, chirality is also exhibited on a higher level by the helices, the tertiary structure, and the folding pattern of most protein molecules [201-203]. Even if one looks at proteins at a low resolution and regards their helical segments as mere rods of some thickness while ignoring the (chiral) internal structures of these helices, most proteins still appear chiral due to the chirality of their tertiary structures.

If the allowed motions, translations, and rotations are restricted to a plane, it then becomes meaningful to consider two-dimensional chirality [37,46-54]. Molecules of planar nuclear arrangements adsorbed on a planar surface of a metal catalyst may be regarded as chemical examples manifesting two-dimensional chirality. If only those motions are allowed which keep the molecular plane parallel with the metal surface, then two molecules, identical in three dimensions, may exhibit planar chirality along the surface. For example, trifluoroethene (achiral in three dimensions) has two possible adsorbed arrangements along a plane: two mirror images which are not superimposable if only those motions (planar translations and rotations) are allowed which preserve their molecular planes parallel with the metal surface. The nuclear framework of the equilibrium configuration of trifluoroethene exhibits two-dimensional chirality.

In both three and two dimensions, chirality is a discrete property: a rigid object is either chiral or achiral. However, a chiral object with a shape that is only marginally different from that of an achiral object may be regarded as being "less chiral" than another chiral object that has very different shape from that of any achiral object. Consequently, it is possible to consider the degree of chirality, that is, it is possible to quantify chirality by numerical chirality measures. Buda, Auf der Heyde, and Mislow have distinguished two types of chirality measures: those which compare a chiral object to some achiral reference object, and those which compare a chiral object to its mirror image [58]. Measures of the former type involve a choice of reference object; a similar approach of comparing objects to some chosen chiral reference object has been advocated by Rassat [47]. Measures of the second type include those described by Kitaigorodskii [46], Gilat [48-50], and

Mislow and co-workers [51-53,57,58]. There are many ways of measuring the geometry differences between a chiral object and its mirror image. One method, introduced by Kitaigorodskii [46], is based on the volume (or area in two dimensions) of the maximum overlap between an object and its mirror image. The ratio q of the volume of the maximum overlap to the volume of the object is one for an achiral object and a smaller value for a chiral object. (Here we assume that the object is nowhere infinitely thin, such as a chiral line segment attached by one end to a ball. Although the entire object is chiral in 3D, the infinitely thin line segment has zero volume, consequently, unless we exclude such pathological infinitely thin cases, q=1 is possible for a chiral object.) A smaller ratio q clearly belongs to an object of more pronounced chirality, consequently, the quantity (1-q) may serve as a formal measure of chirality. A recent important advance has been made by Mislow and co-workers who determined the most chiral constrained and unconstrained simplexes in two dimensions (the most chiral triangles) according to the above criterion [51-53,58].

For more general objects, several chirality measures have been proposed based on the concepts of maximal achiral subsets and minimum achiral supersets [240]. A maximal achiral subset of an object is a subset that cannot be increased within the object without becoming chiral, and a minimal achiral superset of an object cannot be decreased while containing the object and staying achiral. Note that for some objects neither the maximal achiral subset nor the minimal achiral superset is necessarily unique, and their collection gives a fairly detailed chirality characterization [240], for example, by measuring the deviation of their volumes from that of the original object and from one another.

An alternative approach is based on the Hausdorff distance, a formal distance reflecting the differences between two objects [241]. The Hausdorff distance h(A,B) between two sets A and B is the smallest value r such that each ball of radius r centered at any point of either set contains at least one point of the other set. For example, the Hausdorff distance between two superimposed molecular contour surfaces is the smallest r value such that any point on either contour surface has at least one point of the other contour surface within a distance r. The Hausdorff distance h(A,B) is zero if and only if the two sets are the same, A=B. The Hausdorff distance is applicable to measure the deviation of a chiral nuclear arrangement from some arbitrary reference arrangement, as proposed by Rassat [47]. According to the approach of Mislow and co-workers, the Hausdorff distance between the object and the optimally overlapping mirror image provides a chirality measure of the second type [58,242]. Mislow and Buda determined the most chiral constrained and unconstrained simplexes in two and three dimensions (the most chiral triangles and tetrahedra) according to the Hausdorff criterion [242].

The chirality quantification technique proposed by Harary and Mezey [54,55] is motivated by the Resolution Based Similarity Measure (RBSM) approach used in more general molecular similarity analysis [243]. This method does not rely on a single reference object. Instead, it characterizes shape on any desired finite level of resolution by considering various A(J,n) parts of square lattices, called *lattice animals* or P(G,n) parts of cubic lattices called *polycubes* which can be inscribed within the two- or three-dimensional objects J or G, respectively. In the above notation, n is the number of lattice cells in A(J,n) or P(G,n), and A(J,n) or P(G,n) is classified as "interior filling" if no animal or polycube of the same lattice cell size and n+1 cells fits within the object J or G, respectively. Clearly, decreasing the cell size (hence, eventually increasing n) is equivalent to observing the object J or G at a higher level of resolution. The chirality of the interior filling lattice animals or polycubes can be tested by algebraic means for any finite cell number n. The chirality index n_0 of the object J or G is defined as the smallest cell number n at and above which all interior filling animals or polycubes are chiral.

The concept of topological chirality is best illustrated by chain molecules that may form knotted loops or chain links, such as some DNA fragments or some catenanes, respectively. Some of the types of topological molecular chirality can be characterized by methods of knot theory, an important branch of topology. Knots and links may undergo many geometrical changes without changing their "knottedness", (e.g., without changing their topology and fundamental embedding properties within the three-dimensional space) [59-72]. As long as the string of the knot or link is not cut, all possible geometrical rearrangements of a given knot or link lead to an equivalent knot or link, respectively. In spite of this freedom in motions, many (in fact most) knots are chiral: these topologically allowed motions are not sufficient to bring the knot or link into perfect superposition with its mirror image. That is, as long as no bonds are broken, the topologically chiral molecule remains chiral. By interpreting the concept of conformational change in a broad sense as any molecular motion preserving formal chemical bonds, all conformations of a topologically chiral molecule preserve chirality. This freedom in topological chirality is in contrast with a more common occurrence of molecular chirality arising at formal chiral carbon centers, where not all motions preserving the formal bonding pattern preserve chirality. It is possible to force a chiral pyramidal nitrogen center through a planar bond arrangement converting the molecule to its mirror image while preserving the formal bonding pattern between the nitrogen atom and its substituents. Of course, there is still an energy constraint. For example, tetrahedral-to-planar deformations at carbon centers require very high energies, comparable to that required for breaking bonds in a topologically chiral molecule. At all levels, molecular chirality is a function of the available energy.

Whereas some molecules themselves form knots, such as various DNA segments or certain skillfully synthesized chain molecules [59-72], techniques have also been proposed for representing some of the stereochemical properties of arbitrary molecules by knots [62], simply by using knot theory to characterize the space surrounding the molecule. By regarding the molecule as a sculpture, one may think of this approach as characterizing the chirality of the hollow casting shell by a knot [62]. If the molecule is chiral, so is its casting shell. Since various polynomial invariants can be assigned to knots, the stereochemical properties can be described by algebraic methods [244]. These polynomials can be generated independently of the actual geometrical representation of the knot, that is, they depend only on topological "knottedness" properties. Some of these polynomials, for example, the Jones polynomial [244], are suitable to detect topological chirality. More details of topological chirality and these intriguing polynomials are discussed in Chapter 3.

1.3 Point Symmetry Groups and Framework Groups

Symmetry of molecules is one of the most easily recognizable molecular shape characteristics, with important consequences for vibrational behavior, spectroscopic properties, and product distribution in chemical reactions. In colloquial chemical terminology, "molecular symmetry" usually means the point symmetry of a formal geometrical arrangement of the nuclei. Point symmetry groups provide a concise and mathematically precise description of the symmetries of the nuclear frameworks.

The point symmetry groups of molecules represent obvious constraints on molecular shapes. However, symmetry alone provides insufficient direct information for a detailed characterization of the shape of formal molecular bodies: two bodies of the same symmetry may show great differences in their shapes. In fact, with the exception of the spherical symmetry of an isolated atom, for a given symmetry there exists a family of infinitely many possible shapes. Nevertheless, symmetry is a useful tool in shape classification, and interrelations between the nuclear arrangement and the electronic density provide simple rules for the symmetry of formal molecular bodies and contour surfaces. For example, for a molecular species in its electronic ground state, the point symmetry group of the nuclear arrangement is usually a subgroup of the symmetry groups of the various isodensity contour surfaces. Here each group is regarded as one of its subgroups.

In this book we shall place only limited emphasis on symmetry, since there is little use of symmetry in the shape characterization of more complicated molecules most of which have only trivial symmetry. The reader may find many excellent texts on molecular symmetry in the literature (for a selection see references [73-79]). Note, however, that deviations from a given symmetry and various *symmetry deficiency measures* are important and more generally applicable tools for shape characterization. These latter subjects are discussed in Chapter 8.

Whereas symmetry describes only a limited aspect of the shape of a molecule, somewhat more shape information can be deduced from a generalization of molecular point symmetry groups, often called *framework groups*. For each nuclear arrangement, also referred to as nuclear configuration K, the framework group can be specified by the corresponding point symmetry group, combined with information on the location of individual nuclei with respect to the symmetry elements of the configuration and the behavior of the nuclei under the symmetry operations [245]. The information additional to the point symmetry group can also be given by assigning a permutation operator P to each symmetry operator R of the point group according to the following condition: if the permutation operator P is applied after the symmetry operator R, then operator P rearranges all indistinguishable nuclei back to their original arrangement [246]. In other words, the effect of symmetry operator R on the nuclear configuration K is "undone" by the permutation operator P :

$$PR \text{ K} = \text{K}. \tag{1.1}$$

The pairs (P,R) are elements of a group, often called the *framework group* of the nuclear configuration K. The product (P_c,R_c) of group elements (P_a,R_a) and (P_b,R_b) is defined as follows:

$$(P_c,R_c) = (P_a,R_a) \cdot (P_b,R_b) , \qquad (1.2)$$

where

$$R_c = R_a \cdot R_b , \qquad (1.3)$$

and

$$P_c = P_a \cdot P_b . \qquad (1.4)$$

Note that in two different molecules a given common point symmetry operator R can be associated with two different permutation operators P and P', depending on the nuclear arrangements. Consequently, the framework groups do contain more information on molecular shapes than point symmetry groups.

Whereas framework groups are more informative than point symmetry groups, they are able to describe only a rather restricted aspect of molecular shape. Alternative group theoretical methods, notably, the Shape Group Methods (SGM) of molecular topology, are more suitable for a detailed shape characterization. The shape groups will be discussed in Chapter 5 of this book.

1.4 Dynamic Shape Properties: Conformational Freedom and Electronic Excitation

Molecular shape is not a static property. Even at absolute zero temperature, molecules exhibit formal vibrational properties manifested in a probabilistic distribution of nuclear positions in any polyatomic molecule. Rotational states of molecules also influence their shapes. Motion is an inherent property of molecules; consequently, molecular shapes cannot be described in detail without taking into account the dynamic aspects of the motion of various parts of the molecule relative to one another. Within a semiclassical approximation, the dynamic shape variations during vibrations can be modeled by an infinite family of geometrical arrangements. At higher temperatures i.e., if more energy is available, molecular vibrations may cover a wider range of formal molecular geometries, hence a greater variety of dynamic shapes occur. At even higher energies, sufficient for overcoming the activation barriers to formal conformational rearrangements, a further increase in the extent of shape variations can be found. Dynamic shape of molecules is an energy-dependent property.

The conformational freedom of molecules at various temperatures implies a temperature dependence of molecular shapes. For individual molecules, it is meaningful to replace the formal temperature dependence by a dependence on energy: dynamic molecular shape is a function of the energy content of the molecule. For a "cold" molecule that has energy not much exceeding the lowest possible energy content as provided by the zero-point energies of the various vibrational modes, only a limited sample of possible nuclear arrangements (nuclear configurations) are accessible above some probability threshold; hence, only limited shape variations are allowed. Consequently, the dynamic shape of the system is strongly constrained. By contrast, if the molecular system has energy much above the zero-point energy associated with the given potential energy minimum, then the molecule can access a much larger family of possible nuclear configurations with significant probability, hence the dynamic shape is less restricted. The accessible symmetries [247] and the accessible shapes [248] of molecules as a function of the available energy suggest a family of rules influencing the outcomes of chemical reactions.

Another important influence of energy on molecular shapes is evident in electronic excitations. A closed shell molecule with a singlet ground state electronic configuration usually has a different shape than any of the excited electronic states of the same molecule. The potential energy surfaces of different electronic states are usually rather different, their minimum points may occur at different nuclear arrangements, and even the point group symmetries of the most stable nuclear arrangements can be different.

One may formally decompose the molecular shape changes associated with an electronic excitation into separate steps as follows.

1. In an electronic excitation or de-excitation process the electron distribution changes rapidly. However, the nuclear motions are much slower and right after the electronic change the nuclear distribution has had no time yet to rearrange to relax to the neighborhood of the energy minimum of the potential surface of the new electronic state. In the first step of the process the formal nuclear geometry is preserved and the shape of the electron distribution may be regarded as that of the new electronic state at the old nuclear geometry. The shape change of the electron distribution associated with this step of the process is called the *vertical shape change*.

2. In the second formal step of the process, the nuclear arrangement relaxes to a nearby minimum of the potential energy surface of the new electronic state. The three-dimensional body of the electron distribution "follows" the nuclear rearrangement, hence the shape of the electron distribution changes in this step too. This change is called the *shape change due to relaxation*.

In some electronic excitation processes an interesting principle applies, analogous to the Quantum Chemical le Chatelier Principle (QCLCP) originally proposed [249] for explaining regularities found in changes of energy components in various chemical processes. In many chemical processes, e.g., in many conformational changes, various components of the molecular total energy change in the opposite phase. For example, by changing the nuclear arrangement, the nuclear repulsion energy component changes, and this change is often accompanied by a change of opposite sign in the electronic energy component of the total energy. According to the QCLCP, this may be interpreted as an initial stress of nuclear geometry change applied to the molecular system, and the readjustment of the system by changing the electronic energy in the opposite sense, thereby reducing the overall energy change of the molecule. A formal justification of this principle was derived from the variational principle of quantum mechanics [249]. The analogous principle that appears to apply to many shape changes in electronic excitations [111] can be stated as the

Quantum Chemical le Chatelier Principle for Molecular Shapes (QCLCP-MS) : the *shape change due to relaxation tends to reduce the effect of the initial vertical shape change.*

In other words, the initial vertical shape change in Step 1 may be regarded as a stress applied to the molecular system, and the system appears to readjust in Step 2 by the shape change due to relaxation in such a way that the overall shape change is reduced to less than that of the vertical shape change.

A qualitative justification of this principle may be given in terms of energy considerations. Within narrow ranges of energy and electronic density, a monotonic energy change is usually associated with a monotonic change in the complexity of the shape of isodensity contours. A higher-energy electronic distribution well below those involving molecular Rhydberg states usually has a more complicated shape than a lower-energy electron distribution. For example, the nodal structures of higher-energy excited state orbitals (well below Rhydberg orbitals) are usually more complex than those of lower-energy orbitals, suggesting a similar trend for the shapes of isodensity contours. Consequently, the initial vertical shape change in Step 1 is expected to introduce more complicated shape features. However, this initial, vertical shape change is unlikely to result in the most stable nuclear and electron distribution. By relaxation of both the nuclear geometry and the electron distribution in Step 2, the energy is lowered. Energy lowering is often associated with a simplification of shape features; consequently, the shape of electron distribution is also expected to become somewhat simpler. Hence, in Step 2, the shape change is expected to partially counteract the vertical shape change of Step 1, thereby reducing the overall shape change. In general, in any electronic excitation process the initial vertical excitation without a relaxation of the nuclear geometry involves a larger energy change than the overall energy change after relaxation. Consequently, as long as the shape complexity is a monotonic function of the energy, the principle applies.

In the above, qualitative description of the expected trends of interrelations of changes in molecular shape and electronic state, an important element was missing: a quantitative description of molecular shape and a numerical measure of shape changes. A precise, quantitative molecular shape description is also needed in the study of most other problems of chemistry, as well as in various related subjects, such as biochemistry, pharmacology, medicinal chemistry, and drug design.

The intuitive, subjective shape concepts and the freedom provided by a somewhat imprecise shape perception are useful in the process of quick recognition of major trends and dominant common features. Some degree of vagueness in the concept of shape may be advantageous when the goal is to recognize the essential trends and many small details are to be disregarded. However, the needs for clearly defined shape concepts and for precise shape evaluation are evident if details of shape are important, and if it is not well understood which shape features are potentially responsible for a given molecular property. A simple, visual inspection of shape by a human observer may easily miss some important detail, especially, if the relative importance of various details of shape is not yet known. A systematic, precisely defined shape analysis and a detailed and reproducible shape description by a well defined procedure (possibly by a computer algorithm) are preferred.

An ideal molecular shape description method S is expected to fulfill several criteria. An ideal method S

1. is based on the physical properties of the molecule,
2. describes the full, three-dimensional shape of the molecule,
3. leads to numerical shape characterization, such as a numerical shape code,
4. is easily computable, leading to computer-based molecular shape analysis,
5. is reproducible, not affected by subjective elements of human perception,
6. provides tools for the evaluation of shape similarity,
7. provides tools for the evaluation of shape complementarity.

In the following chapters we shall discuss the fundamental physical basis of the molecular shape concept, and describe several of the computational methods of molecular shape description, fulfilling some or all of the above criteria.

CHAPTER
2

THE QUANTUM CHEMICAL CONCEPT OF
MOLECULAR SHAPE

In this chapter some of the theoretical, quantum chemical, and computational aspects of molecular shape are discussed. The quantum chemical shape of molecules is intimately related to the concepts of nuclear configuration and chemical bonding. Both of these concepts can be reformulated in terms of topology. A topological representation of nuclear configurations provides a systematic treatment of the nuclear configuration space and the catchment regions of potential surfaces, leading to a description of molecular deformations which preserve chemical identity. It is the "glue" of the fuzzy, three-dimensional body of electronic density that holds molecules together; chemical bonding is not restricted to formal bonding lines between atomic components of a molecule. Accordingly, topology also gives a descriptive method and new insight for the representation of chemical bonding in terms of the topological Density Domain Approach (DDA), and a quantum chemical definition of functional groups within molecules.

2.1   Electron Distributions and Nuclear Distributions: the Heisenberg Uncertainty Relation and Molecular Shape

The usual image of a molecule invoked in contemporary chemistry is a curious combination of quantum mechanical and classical mechanical models. Whereas it is well accepted that even a crude description of the electron distribution within a molecule must rely on quantum mechanics leading, for example, to various molecular orbital approaches, the nuclear distribution, in contrast, is most commonly imagined as a classical mechanical arrangement of tiny particles. This traditional molecular model persists in spite of the failures of the particle model, for example, when molecular vibrations are considered where the nuclear motions have clear quantum mechanical character as evidenced by the quantized vibrational energy levels. Nevertheless, chemical thinking is still guided by the classical model of nuclear arrangements and few chemists imagine nuclei as formal positive charge clouds analogous to the image of negative and more diffuse charge clouds of electron distributions. On an intuitive level, electrons of a molecule are regarded quantum mechanically, whereas nuclei are often treated classical mechanically. Whereas the above mixed model is wrong by the standards of rigorous quantum mechanics, this formal lack of consistency nevertheless has a well justified, practical motivation that can be phrased in terms of the Heisenberg uncertainty relation.

Molecular shape is the shape of the electron distribution of the molecule. This electron distribution can be described according to the laws of quantum mechanics and within a molecule the wave nature of electrons seems to dominate over their particle-like properties. On the other hand, for atomic nuclei in a molecule, the particle-like properties often appear dominant over their wave-like properties. The concept of the position of an electron within a molecule is rather meaningless. In contrast, the concept of nuclear position within a molecule may serve as a crude, but still useful, approximation. However, even when using this approximation, for example, in the most common form of the Born-Oppenheimer approximation [250], one should remember that both electrons and nuclei obey quantum mechanics. In a strict sense, just as electrons, nuclear arrangements are also subject to the Heisenberg uncertainty relation and it is just as misleading to consider precise relative positions for nuclei as it is wrong to assume precise positions for electrons within a molecule. Hence, nuclear position in a molecule is also a somewhat artificial concept. As a consequence of these uncertainties in position, there are two reasons why the body of a cloud-like fuzzy electron distribution of a molecule is very different from any macroscopic body [251,252]. A minor, indirect contribution to this fuzziness is due to the relatively small quantum mechanical uncertainty in the nuclear positions controlling the electronic distribution, and a major, direct contribution is due to the electron distribution having a much greater quantum mechanical uncertainty of its own. Within a rigorous quantum mechanical description, the above distinction of various contributions to the fuzziness of molecular bodies is somewhat artificial; nevertheless, there are valid practical reasons for it. Nuclei have much larger masses than electrons, they are more particle-like than electrons, and, as a result, the particle model is a much better approximation for their description than for the description of the electronic distribution. In many molecular modeling applications the concept of nuclear position is acceptable for representing reality to a good approximation, whereas the notion of electronic position within a molecule is clearly unacceptable in nearly all instances.

Hence, a simple, essentially classical model provides a useful approximation to the relations between the electronic and nuclear distributions: one may think of the electron distribution as a formal charge cloud, and the nuclear distribution as an arrangement of particle-like nuclei within this charge cloud. The electron distribution is influenced by the nuclear distribution and by external electromagnetic fields; in fact, the shape of the electronic cloud is influenced by all the interactions affecting the molecule. The model is the simplest if there are no external fields. The nuclei may be thought of as point-like objects moving in the field of the electron distribution that can nearly instantaneously readjust to follow the changes of the nuclear distribution. Hence, the shape of the electronic cloud does reflect the nuclear arrangement, and in most traditional stereochemical approaches the concept of molecular shape has been interpreted in terms of the nuclear arrangement. If, however, external fields are applied, then the electron distribution may undergo changes that may result in a dramatic shape change, affecting both the electronic and the nuclear distributions. Since the electron distribution is much more mobile than the nuclear arrangement, the initial stages of these rearrangements are usually dominated by a rapid rearrangement of the electron distribution. The old nuclear distribution is no longer favorable for the electronic arrangement, and a slower nuclear rearrangement follows, accompanied by further, nearly instantaneous, electronic rearrangement in response to the combined effects of the external field and the change in the nuclear distribution. Eventually, a new nuclear and electronic distribution is established that is compatible with the external field. One extreme example is the external field provided by another, interacting, molecule: the interaction may result in a chemical reaction changing the shapes of both molecules.

2.2 The Concept of Topological Shape of Molecules

Molecules are dynamic objects undergoing continuous internal motion. Some finite range of possible deformations with respect to the formal equilibrium shape of the molecule is an inseparable aspect of any realistic molecular model. Consequently, it is important to use techniques for molecular shape characterization which can account for the deformability and the dynamic features of molecular shapes. One must be able to distinguish the essential shape deformations from those having little chemical significance. For example, most small molecular deformations do not change the chemical identity of the molecule, however, extensive deformations may lead to dissociation or to some other chemical reaction that changes the chemical identity of the species. For consistency in the description, we shall associate a separate chemical identity with each stable conformer of a molecule; indeed, as long as an energy barrier separates two conformers, similar considerations apply as for a reactant and product pair of molecules separated by an energy barrier. There is a finite range of possible deformations which preserve the chemical identity of the conformer.

Two important questions of molecular shape analysis are as follows:

1. What is the range of deformations which preserves chemical identity, and
2. What shape variations accompany these identity-preserving molecular deformations?

For the first question, a topological analysis of molecular potential energy surfaces can provide an answer, leading to the concept of *catchment regions* ([106] and references therein). Below we shall give a brief introduction to the global topological analysis of chemical identity preserving deformations, using the concepts of the *nuclear configuration space, potential energy surfaces,* and *catchment regions.* A topological study of potential surfaces can be combined with a three-dimensional topological analysis of molecular shapes, leading to a systematic approach to the second question.

Within a global approach to the study of molecular deformations and their relations to molecular identity, it is advantageous to use the *nuclear configuration space* approach. The simplest, nontrivial nuclear configuration space is that of a diatomic molecule. For diatomics there is only one internal coordinate: the most natural choice for this coordinate is the internuclear distance d. If all the possible values of internuclear distances are plotted along a line, then each point of this line represents a nuclear arrangement. This line is a one-dimensional configuration space, representing the entire family of all possible arrangements of the two nuclei. In fact, only a half-line is used, since no negative internuclear distances are possible. For more complicated molecules the description of the nuclear arrangements requires more than one internal coordinate (for example, several internuclear distances, bond angles, and dihedral angles), hence the line must be replaced by some higher-dimensional space, called the nuclear configuration space. One may picture this space as follows: each point of this space represents a nuclear arrangement so that points corresponding to similar arrangements are near one another in the space. A nuclear *rearrangement* corresponds to a *path* in this space.

Here we shall justify the nuclear configuration space approach within the context of the simple semiclassical molecular model of essentially particle-like nuclei embedded in a quantum mechanical electronic cloud, as described in the previous section. (One should note, however, that the nuclear configuration space model can be introduced on a more rigorous quantum mechanical level [106].) Consider all possible arrangements of a given family of nuclei, called a *stoichiometric family of nuclei.* Each *stoichiometric family of chemical species* is defined by such a set of nuclei which may take *any relative arrangement.* These arrangements include those of all distorted forms of conformations, all isomers, reaction intermediates, transition structures, and decomposition products of all molecules with the given atomic composition. The collection of all these arrangements is a *stoichiometric family of nuclear configurations.*

The identity of a molecule, far removed from other molecules and from sources of external fields, does not depend on its precise location and orientation in the ordinary, three-dimensional space. Consequently, one may regard nuclear arrangements obtained from one another by rigid translation and rigid rotation as chemically equivalent, and consider only the *relative arrangement of the nuclei with respect to one another,* that is, the *internal configuration* K of the arrangement. The family of all possible internal configurations of a set of nuclei may be thought to form an abstract space, the internal configuration space. Each nuclear arrangement corresponds to a point K of this space. Two arrangements which differ only slightly, correspond to a pair of points K and K' which are near each other within this space. This idea can be made precise by introducing a proper distance (called metric) in this space: the configuration space we shall use is the metric space M of internal configurations [106]. The metric of the nuclear configuration space M is interpreted as the distance d(K,K') between any two points K and K' of the space M, representing a *measure of dissimilarity of the corresponding two internal nuclear configurations* K and K'. If the distance d(K,K') is large, then the two nuclear arrangements K and K' are very dissimilar; if d(K,K') is small, then K and K' are similar. Note that we use the same notation K for the *internal configuration* (the 3D relative arrangement of the nuclei) and for the *point* representing it within the configuration space M. Since we disregard rigid translations and rotations of the molecule as a whole, each relative arrangement K of N≥3 nuclei can be described by 3N-6 *internal coordinates.* That is, for larger than diatomic systems the dimension of the nuclear configuration space M is 3N-6, and each point K of M can be described by 3N-6 (local) coordinates.

For example, for the hydrogen peroxide molecule, H_2O_2, there are four nuclei, hence 3 x 4 - 6 = 6 internal coordinates. A unique label is assigned to each nucleus, for example, we may choose an assignment H_a - O_b - O_c - H_d. For most nuclear arrangements of hydrogen peroxide, the six internal coordinates can be chosen as the three internuclear distances H_a - O_b, O_b - O_c, and O_c - H_d, the two bond angles H_a - O_b - O_c and O_b - O_c - H_d, and the dihedral angle denoted by H_a - O_b - O_c - H_d, i.e., the angle between the planes defined by the first three and the last three nuclei in the sequence H_a - O_b - O_c - H_d. One should note that these internal coordinates are not well defined for all nuclear arrangements, for example, the dihedral angle H_a - O_b - O_c - H_d is not defined if either the H_a - O_b - O_c or the O_b - O_c - H_d triple of nuclei become colinear. Nevertheless, it is possible to describe even such nuclear configurations within a consistent framework [106], and the collection of all possible nuclear arrangements K of the entire H_2O_2 stoichiometric family of chemical species forms a six-dimensional nuclear configuration space M.

Most deformations of objects involve a change of their energy. The problems of molecular deformations are also related to energetic properties. The study of molecular identity requires the explicit consideration of the energy dependence of the deformability of molecules. This requirement naturally leads to the *molecular potential energy surface model* and to the global analysis of deformability of a whole range of formal nuclear arrangements. For the simplest, nontrivial case of diatomics, the potential surface becomes a one-dimensional potential curve, that may be plotted above the one-dimensional configuration space: the line of the coordinate of the internuclear distance of the diatomic molecule.

The more general, higher-dimensional potential energy surface model is motivated by a simple analogy with 2D surfaces of 3D objects. By considering a given electronic state and the model of infinitely slow motion for the nuclei, a potential energy value can be assigned to each nuclear configuration K; this defines a potential energy function called the potential energy surface E(K) over M. Since the configuration space M usually has a high dimensionality (3N-6 coordinates) and energy can be taken as a formal coordinate along one additional dimension, this surface is often referred to as the *potential energy hypersurface* E(K) [106].

Consider again the example of H_2O_2 of six internal coordinates, as discussed above. For a given electronic state of the hydrogen peroxide molecule (in fact, for the entire H_2O_2 stoichiometric family of chemical species), a potential energy value can be assigned to each nuclear configuration K, and the resulting potential energy function E(K) can be thought of as a potential energy surface (a six-dimensional hypersurface) spanned above the six-dimensional nuclear configuration space M. Energy may be thought of as the seventh coordinate, and the six-dimensional energy hypersurface E(K) as being embedded in a seven-dimensional space. Of course, for a larger molecule of more than four nuclei, N > 4, the dimension 3N-6 of the configuration space M and the hypersurface E(K) is also higher; an additional coordinate, the (3N-5)-th coordinate corresponds to energy. The corresponding (3N-6)-dimensional potential energy hypersurface E(K) is embedded in a (3N-5)-dimensional space.

Most of the ordinary notions of surfaces in three dimensions can be generalized for such hypersurfaces. For example, the slope of a surface at each point can be described by its partial derivatives, and the slope of the potential energy hypersurface at each nuclear configuration K can be described by the partial derivatives of energy according to the internal coordinates, i.e., by the gradient of E(K) at this point K. This gradient is a formal (3N-6)-dimensional vector in the nuclear configuration space M. The slope of the energy hypersurface, (i.e., the negative of the gradient vector) represents a formal force within the nuclear configuation space M. This formal force acts on the configuration K, forcing it to change into some nearby lower energy configuration K' found along this force vector.

A point K of M where the gradient of E(K) vanishes [where the tangent hyperplane to E(K) is "horizontal"], is a point where the force of deformation is zero, i.e., point K represents an *equilibrium configuration.* Such a point is called a *critical point,* and is denoted by $K(\lambda,i)$. Here, the first derivatives being zero, the second partial derivatives of the energy hypersurface are used to characterize the critical points. The first quantity in the parentheses, λ, is the *critical point index* (and not the "order of critical point" as it is sometimes incorrectly called). The index λ of a critical point is defined as the number of negative eigenvalues of the Hessian matrix $H(K(\lambda,i))$, defined by the elements

$$H_{jk}(K(\lambda,i)) = \partial^2 E(K(\lambda,i))/\partial q_j \partial q_k \qquad\qquad (2.1)$$

of second partial derivatives of the energy function E(K) at $K(\lambda,i)$, whereas the second quantity i in the parenthetical expression is simply a serial index.

Each electronic state of the given stoichiometric family corresponds to a formal potential energy hypersurface E(K). Clearly, the notions of chemical species, chemical identity, and molecular deformation are dependent on the electronic state. A specific nuclear arrangement that is stable for one electronic state may be unstable for another. For a given potential energy hypersurface E(K), the range of deformations that preserves chemical identity of a chemical species defines a collection of nuclear configurations, which form a subset of the nuclear configuration space M. This subset is called a *catchment region* $C(\lambda,i)$ of the corresponding potential energy hypersurface. Each catchment region can be taken to represent the given chemical species within M. A catchment region $C(\lambda,i)$ is defined as the collection of all those nuclear configurations K from where the path of an infinitely slow relaxation (a steepest descent path in a mass-weighted coordinate system)   leads to a common critical point $K(\lambda,i)$, representing an equilibrium nuclear arrangement. (See original references in [106].) A *catchment region $C(0,i)$ of a minimum point $K(0,i)$ of E(K) represents the i-th stable molecular species* of the given stoichiometry and of the electronic state associated with the given potential energy hypersurface E(K). The steepest descent paths from all points of the catchment region $C(0,i)$ lead to the unique minimum point $K(0,i)$; the dimension of $C(0,i)$ is 3N-6. A (3N-7)-dimensional *catchment region $C(1,j)$ of a saddle point $K(1,j)$* of critical point index $\lambda=1$ *represents the j-th transition structure* (transition state as it is often incorrectly called). Catchment regions $C(\lambda,i)$ of critical points $K(\lambda,i)$ of higher indices, $\lambda > 1$, and of dimensions lower than 3N-7, are of lesser direct chemical significance (and are not true chemical species), nevertheless, for the sake of uniformity in the terminology, they are also referred to as formal chemical species. In particular, we shall not consider the case of potential energy maxima, which are single-point catchment regions C(3N-6,i) (also called singleton sets).

The above catchment region approach provides a general basis for the study of the interrelations of molecular nuclear arrangements and 3D molecular shape [158], leading to a common framework for the description of a variety of molecular shape problems, ranging from molecular similarity to shape changes in electronic excitations [107-111]. Through the potential energy hypersurface approach, the three-dimensional molecular shape problem is connected with   many other energy-dependent chemical problems. Electronic and vibrational properties, conformational freedom, reactivity, bond formation, and bond breaking are all energy-dependent, and the potential energy surface approach provides an elegant, conceptually convenient, although rather complicated (multidimensional) representation of this energy dependence. The topological analysis of potential energy surfaces in terms of catchment regions simplifies this representation and provides a unified framework for the study of individual molecular properties, all conformational changes, as well as chemical reactions [106]. The three-dimensional topological properties of molecular shape are intimately connected with the (3N-6)-dimensional topological properties of potential energy hypersurfaces.

The catchment region model provides precise conditions for chemical identity and for limitations on molecular distortions which preserve chemical identity. It also provides an approach to our second question: What are the allowed shape variations which may accompany these identity-preserving molecular deformations?

When considering the shapes of various distorted forms of a molecule, one must make a choice of a reference form against which all other forms are compared.

In a classical model of molecules, with some geometrically defined ideal form, all other forms may be compared to the ideal one. For example, the results of experimental X-ray structure determination are often interpreted in terms of a simple classical model of formal, point-like nuclear positions within a molecule and the point arrangement of this interpretation may be regarded as a molecular reference form. Even if one disregards the fact that molecular shape in the solid state can be very different from the shape of a free molecule, the above approach is limited to one or a few arrangements. The global topological analysis of potential surfaces provides a more detailed alternative. There is a one-to-one correspondence between critical points $K(\lambda,i)$ and catchment regions $C(\lambda,i)$ of potential energy surfaces, which "cover" the entire nuclear configuration space M (i.e., which involve all possible nuclear arrangements of the stoichiometric family). Each critical point $K(\lambda,i)$ represents the equilibrium nuclear configuration of the corresponding catchment region $C(\lambda,i)$, and $K(\lambda,i)$ may be used as reference for distortions within each catchment region $C(\lambda,i)$.

The family of 3D shapes available to a given molecule is precisely the family of 3D shapes occurring within its catchment region $C(\lambda,i)$ [158]. These are the very shapes attainable by the molecule while undergoing limited deformations preserving chemical identity.

2.3 Molecular Isodensity Contours (MIDCO's)

For sake of simplicity in describing the essential ideas within this and the following sections, first we shall consider a simplified model of a formal, static nuclear arrangement for each molecule. This constraint on the model will be released when dynamic shape analysis is discussed.

For a given electronic state, the nuclear arrangement K is the dominant factor that determines the shape of the formal body of molecular electron distribution. External electromagnetic fields also influence molecular shape, nevertheless, their effect is usually only secondary. Whereas the nuclear configuration K contains most of the information necessary for shape analysis, the dependence of the electron distribution on K is rather complicated, and it is not immediately obvious what molecular shape can be expected for a given nuclear arrangement. For this reason, it is advantageous to study the shapes of electronic density distributions directly.

An important tool for the description of 3D electron densities is the concept of *molecular isodensity contours* (MIDCO's). For any formal nuclear configuration K of a molecule we may assume a 3D coordinate system attached to the chemical species. Within this coordinate system, the electronic charge density $\rho(r)$ is a function of the 3D position variable r; this function assigns a density value $\rho(r)$ to each point r of the ordinary 3D space. This charge density function $\rho(r)$ can be determined experimentally, for example, from X-ray diffraction experiments [89,90], or it can be calculated by an appropriate quantum chemical method, for example, using one of the standard quantum chemistry programs [253] or some approximate technique designed for large molecules [92]. Since the physical properties of the peripheral regions of molecules are dominated by the electronic density, it is natural to associate the concept of molecular shape with the regions of space enclosing most of the electronic charge density of molecules.

Whereas the charge density function $\rho(\mathbf{r})$ becomes zero only at infinite distance from the nuclei of the molecule, this function converges rapidly to zero already at short distances, becoming negligible at about 10 Å from the nearest nucleus. In fact, the electronic density is well localized within a close neighborhood of the nuclei and it is justified to regard only those regions of the 3D space as belonging to the molecule where the density $\rho(\mathbf{r})$ is larger than some small threshold value. By choosing a sufficiently small threshold value a, an approximate molecular *body* can be defined as the collection F(a) of all those points $\mathbf{r}$ of the 3D space where the electronic density is greater than the threshold a,

$$F(a) = \{\ \mathbf{r} : \rho(\mathbf{r}) > a\ \} . \tag{2.2}$$

In equation (2.2) some of the standard mathematical notations are used; these notations are very useful for a precise and concise expression of ideas we shall describe. The pair of braces { } stands for "collection" or "set", whereas the colon : stands for the qualification "such that" ; the expression reads as "the collection of all points $\mathbf{r}$ such that $\rho(\mathbf{r}) > a$ ". Using mathematical terminology, these points $\mathbf{r}$ are said to form a *level set* F(a), with respect to the threshold level a. According to the usual convention, a large *negative* charge means a large *positive* value for the density function $\rho(\mathbf{r})$. Of course, the size and shape of such a level set F(a) depend on the choice of the threshold value a. We shall not select any arbitrary value for a, instead we shall study a whole *range* of possible a values for electronic density level sets. In principle, the range for the threshold parameter a is $0 \leq a \leq \infty$, however, in molecular shape analysis only a more restricted range of the chemically relevant density thresholds will be considered. It is clearly not necessary to consider very high a values. Furthermore, in order to avoid considering a common level set with a single, common envelope surface for independent molecules of large but finite distance from one another, a small but nonzero lower limit will be taken for the a values.

One should notice that according to the definition (2.2), those points $\mathbf{r}$ of the 3D space where the value of the electronic density function is equal to the threshold,

$$\rho(\mathbf{r}) = a, \tag{2.3}$$

do not belong to the level set F(a). This is a matter of definition, and for some purposes it is advantageous to adopt an alternative definition by including all points $\mathbf{r}$ of the space where equation (2.3) holds, that is, all the points *and all the boundary points* of the level sets F(a) of definition (2.2). Such domains of the 3D space, closely related to the level sets F(a), are of special importance since they provide a natural representation of chemical bonding in molecules. These domains, the *density domains* [109] are denoted by DD(a) and will be discussed in the next section, Chapter 2.4.

It is useful to emphasize that the above representation of a formal molecular body by a *single* level set F(a) and a *single* nuclear arrangement K involves two simplifications [109]:

1. It is assumed that a fixed nuclear configuration K adequately represents the actual quantum mechanical distribution of the nuclei.

2. The model refers to a specified density value a along the contour. Those points of space where the electronic density is less than this threshold value a are not regarded to belong to the formal molecular body considered.

Both of these constraints can be released in a more general model, where a *family* of level sets and a *family* of nuclear arrangements are considered simultaneously.

Level sets F(a) [as well as the closely related density domains DD(a), as we shall see in the next section] provide a representation of *formal molecular bodies*. A similar definition gives a useful concept of a *formal molecular surface:* the concept *molecular isodensity contour surface* (MIDCO). For any formal nuclear configuration K, it is possible to define a surface by choosing a small value a for the electronic density, and by selecting all those points $\mathbf{r}$ in the 3D space where the density $\rho(\mathbf{r})$ happens to be equal to this value a, that is, where equation (2.3) is fulfilled. For an appropriate small value a, this contour surface may be regarded as the surface of the essential part of the molecule and, in short, it may be referred to as the molecular surface. These surfaces, the *molecular isodensity contour surfaces,* or MIDCO's, are denoted by G(a) and are defined as

$$G(a) = \{ \ \mathbf{r} : \ \rho(\mathbf{r}) = a \ \}. \tag{2.4}$$

For a continuous function, such as the electronic density $\rho(\mathbf{r})$, all points $\mathbf{r}$ fulfilling equation (2.3) do form a continuous surface. Consequently, the terms *contour surface* and *isodensity surface* are appropriate for G(a). For the study of the 3D shape properties of molecular bodies, represented by level sets F(a) of electronic charge densities, it is sufficient to study the shape of their boundaries; these boundaries are the MIDCO's G(a).

In Figure 2.1 four MIDCO's, $G(a_1)$, $G(a_2)$, $G(a_3)$, and $G(a_4)$ of the methanol molecule CH_3OH are shown for the contour density values $a_1 = 0.20$, $a_2 = 0.10$, $a_3 = 0.01$, and $a_4 = 0.001$, respectively, as calculated with the GAUSSIAN 90 [253] and GSHAPE 90 [254] programs, using a 6-31G* Gaussian basis set. The first three MIDCO's show characteristic shape features; the methyl and hydroxyl groups are clearly distinguishable in the first two MIDCO's, and even in the third, bulkier, MIDCO $G(a_3)$ of $a_3 = 0.01$, the orientation of the OH group is well recognizable. The lowest density MIDCO at $a_4 = 0.001$ shows a rather general feature of low density surfaces: they become essentially spherical, where the details of functional groups of the molecule are no longer recognizable. One might ask the question, which one of the above four MIDCO's is the *true* representation of the methanol molecule? The answer is that by itself not any one of them is, and in principle, one needs the entire collection of all possible MIDCO's for the entire range $(0, a_{max})$ of contour density values a

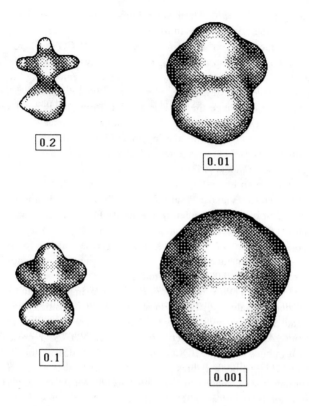

Figure 2.1 Four MIDCO's, $G(a_1)$, $G(a_2)$, $G(a_3)$, and $G(a_4)$ of the methanol molecule CH_3OH are shown for the contour density values $a_1 = 0.20$, $a_2 = 0.10$, $a_3 = 0.01$, and $a_4 = 0.001$, respectively, as calculated with the GAUSSIAN 90 [253] and GSHAPE 90 [254] programs, using a 6-31G* Gaussian basis set. In all figures, the density threshold values a are given in atomic units.

in order to have an exhaustive description of the shape of the molecule. Fortunately, a sufficiently precise shape description is possible by considering only a *finite number* of density intervals $[a_i, a_{i+1}]$ instead of the infinitely many different MIDCO's separately, where, within each such interval $[a_i, a_{i+1}]$, the associated $G(a)$ MIDCO's preserve their essential shape features. This idea, combined with an algebraic-topological description of the essential shape features, forms the basis of the Shape Group Method (SGM) [155-158], discussed in later chapters of this book.

It is possible to generalize the MIDCO model and to release the constraints represented by simplifications 1 and 2. By taking a *collection* of nuclear configurations K, as well as a *range* of small threshold values a, one may generate a family of formal molecular bodies and investigate their *common* shape features. These common features can be described by topology, resulting in a topological definition of the molecular body. In the most general case the geometrical characterization of individual contour surfaces is replaced by a topological characterization of entire families of MIDCO's, where each family belongs to a density interval $[a_i, a_{i+1}]$ and to a selected family of internal configurations K, for example to a catchment region $C(\lambda,i)$. For the corresponding *family* of MIDCO's the *common topological features* provide a valid characterization. Since the nuclear configuration K is not fixed within a catchment region $C(\lambda,i)$, the resulting description involves some dynamic properties and at least partially circumvents the incompatibility of the more common static models with the Heisenberg uncertainty relation. Consequently, this approach provides a more valid model [158].

One can make an interesting comparison between MIDCO's and the simpler models of fused sphere Van der Waals surfaces (VDWS's) often used in modeling large molecules of biochemical interest. For most of the common choices of formal atomic Van der Waals radii, for example, those suggested by Gavezotti [86], the resulting fused sphere VDWS shows a strong resemblance to an electronic isodensity contour surface G(a) of some intermediate threshold value a, approximately equal to 0.002 a.u. (atomic units). One may exploit this fact and design fused spheres VDWS representations of approximate molecular surfaces for any desired density value a, based on atomic charge densities. It is possible to generate a function of density-dependent radius $r_A(a)$ for a spherical representation of each atom A, and to construct a fused sphere VDWS representation of a molecule that mimics a G(a) MIDCO for *any selected* density threshold value a. By choosing a desired density threshold value a, and calculating the appropriate $r_A(a)$ set of radii for the spheres of each atom A, the fused spheres model provides an approximation of the G(a) MIDCO of the molecule for this a value [255]. In addition, for molecular families of common structural features, or for typical functional groups, it is possible to choose an optimum set of atomic VDW radii, based on molecular contour surfaces of electronic charge densities [255].

2.4 The Density Domain Approach (DDA) to Chemical Bonding

Formal chemical bonds of a molecule are usually imagined as lines interconnecting the nuclei, providing a simple model for a molecular skeleton. This model is certainly an oversimplification, since a bond skeleton cannot fully represent the molecular body and the shape of the molecule. In reality, molecules are three-dimensional objects held together by fuzzy bodies of electronic charge distributions. The actual chemical bonding is not constrained to a set of lines in space, neither are the interactions between parts of the molecule concentrated along some narrow bonding channel. Chemical bonding involves a broad interfacing of molecular fragments. Consequently, a more appropriate description of chemical bonding must take into account the full molecular body. Since the molecular body is a 3D object, it appears natural to search for models where chemical bonding is also represented by 3D entities, such as the actual 3D electron distribution that holds the molecule together. Electronic charge densities, regarded as 3D bodies, their level sets F(a) and their isodensity contour surfaces G(a) offer an alternative to the classical stereochemical bond diagrams, and a more realistic model for the representation of chemical bonding.

The study of 3D electronic densities has become a relatively simple task as a result of advances in computational quantum chemistry, molecular topology, and computer graphics techniques. Graphical information on calculated 3D molecular properties can be displayed and manipulated easily on a computer screen, providing a powerful tool for the study of stereochemistry, molecular shape, reaction mechanisms, and the evolution of chemical reactions. Whereas 3D representations and computer plots of individual localized orbitals [256-260] are often invoked in a qualitative rationalization of molecular properties and reactions, the concepts needed to describe the true 3D nature of bonding in molecules, as represented by formal bodies of electronic charge density clouds, are slow to replace the stereochemical ball-and-stick bonding models. The experimental chemist's concept of the chemical bond is still rather traditional, it has not yet exploited the potential for evolution offered by modern quantum chemistry and computer graphics methods. Modern computer graphics techniques are suitable to describe far more complex 3D patterns than simple reincarnations of ball-and-stick models on the computer screen. These advances make it practical to introduce models where chemical bonding is described by more realistic approaches, taking into account the full 3D body of molecular electron distributions.

The simple model of bonds between atoms, reducing chemical bonding to formal atom pair interactions is unsatisfactory for many molecules, since it fails to represent the actual bonding in conjugated or aromatic systems. In reality, chemical bonding is a molecular property, not a property of atom pairs.

The Density Domain Approach (DDA) to chemical bonding has been proposed [109] as a tool that is able to describe the global properties as well as the fine details of the full, three-dimensional bonding pattern within molecular bodies.

The concept of density domains is related to the concept of MIDCO in a simple way. A maximum connected part of an isodensity contour surface G(a) and the corresponding part of the level set F(a) enclosed by it is called a *density domain,* $DD_i(a)$ [109]. Below we shall give a more formal definition and describe the most essential properties of density domains.

A level set F(a), as defined by equation (2.2), is an *open* set; F(a) does not contain its boundary G(a). A *closed* variant of a level set can be defined as

$$DD(a) = \{\ \mathbf{r} :\ \rho(\mathbf{r}) \geq a\ \}\ .\qquad\qquad(2.5)$$

This set DD(a) contains its boundary, in fact, the set DD(a) is the union of the level set F(a) and the boundary G(a) of F(a),

$$DD(a) = F(a)\ \cup\ G(a).\qquad\qquad(2.6)$$

Here the union symbol $\cup$ is used for indicating that the two families of points,

F(a) and G(a), are now considered as a single family, denoted by DD(a). A DD(a) set can be regarded as the *molecular body at an electronic density threshold* a. Indeed, if we imagine that we can actually see molecules, and the sensitivity of our eyes is adjustable to notice only electronic densities that are equal to or greater than the value a, we would then see these DD(a) sets as the molecular bodies.

A DD(a) body may be a single piece or it may be a collection of several disconnected parts, called *maximum connected components*, and denoted by $DD_i(a)$. In general, one may write

$$DD(a) = \bigcup_i DD_i(a), \qquad\qquad (2.7)$$

since every DD(a) body must be the union of its maximum connected components $DD_i(a)$. If DD(a) consists of a single piece, then the DD(a) body is the same as its single maximum connected component $DD_1(a)$,

$$DD(a) = DD_1(a). \qquad\qquad (2.8)$$

The $DD_i(a)$ maximum components of the DD(a) body are the *density domains* of the molecule at density threshold value a. As follows from their definition, their individual contour surfaces, denoted by $G_i(a)$, are also the maximum connected components of the MIDCO G(a):

$$G(a) = \bigcup_i G_i(a), \qquad\qquad (2.9)$$

In fact, the same shape information can be deduced by studying the $DD_i(a)$ density domains or their respective boundary surfaces $G_i(a)$.

The Density Domain Approach to chemical bonding is based on the topological analysis of the dominant shape variations of the molecular body DD(a) [or, equivalently, those of the G(a) contour surface] regarded as a function of the density parameter a.

It is possible to follow the shape changes of the body DD(a) and its density domains $DD_i(a)$ as the density threshold parameter value a scans the interval (0, ∞). In fact, only a chemically significant subinterval $[a_{min}, a_{max}]$ is to be scanned. Very high densities do not occur in most molecules and even those high densities which do occur have little direct chemical relevance as they represent the core regions near the nuclei where the changes are negligible in most chemical processes. Consequently, the upper limit ∞ can be replaced by a finite positive value a_{max}. Similarly, points **r** of the space with very low density values, which occur far from the nuclei, have little direct chemical importance. Furthermore, the chemical interpretation of level sets with near zero density thresholds can become misleading and artificial; as an extreme case, the formal density domain of the a=0 threshold is the whole universe. In order to avoid considering common density domains for independent molecules of large but finite distance from one another, a small positive lower limit a_{min} is taken. According to the convention adopted, we scan the interval $[a_{min}, a_{max}]$ from the right to left, that is, from high to low density values.

At a chemically relevant high electronic density threshold value a, the body DD(a) is composed of several disconnected, nearly spherical density domains $DD_i(a)$. At such high density values a each separate component $DD_i(a)$ of DD(a) contains one nucleus. As the contour parameter a decreases, the isodensity contours $G_i(a)$ expand and various parts of the contour surface G(a) become connected, [i.e., the corresponding density domains $DD_i(a)$ unite to form a single, new density domain]. The sequence according to which various parts of the body DD(a) become connected provides information on the 3D shape of the molecular electronic charge density, indicating the pattern by which the electron density bonds the molecular fragments together. Some molecular fragments retain separate density domains within a wide interval of density values a; usually, these fragments have a well established chemical identity as a "functional group". A gradual decrease of the density threshold value a is accompanied by a characteristic sequence of topological changes of density domains $DD_i(a)$, which eventually leads to the interconnection of all parts of the body DD(a). The isodensity contour G(a) becomes a single envelope surface, surrounding all the nuclei of the molecule, [i.e., the body DD(a) becomes a single density domain $DD_1(a)$]. For the extreme case of very small isodensity contour values a, the surface G(a) becomes a nearly spherical balloon and the body DD(a) becomes a single ball.

The most essential, topologically significant changes are the connections between the density domains $DD_i(a)$. *These changes occur only at a small, finite number of selected density threshold values* a_j. The shapes of various $DD_i(a)$ parts of DD(a) and the pattern of their connections, as a function of the parameter a, provide a new, systematic approach to 3D chemical bonding. A topological description of chemical bonding is given by the finite sequence of families of density domains

$$\{DD_i(a_1)\},\ \ \{DD_i(a_2)\},\ \ \dots\ \ \{DD_i(a_j)\},\ \ \dots\ \ \{DD_i(a_k)\}, \qquad (2.10)$$

as they occur while the density threshold value is gradually decreased within a chemically relevant density range [0, a_{max}], where any two families that are neighbors in the sequence are topologically different.

We call such a sequence a *topological sequence of families of density domains* or, since each family $\{DD_i(a_j)\}$ represents a formal molecular body $DD(a_j)$ at the density level $a=a_j$, a *topological sequence of molecular bodies.* Below we shall discuss several examples as illustrations of the DD approach. The techniques of actual topological characterization will be described in the following chapters.

One may summarize the essence of the approach as follows. The Density Domain Approach is a 3D *topological tool for a comprehensive description of chemical bonding* [109]. By decreasing the contour parameter a from high values to zero, the various density domains $DD_i(a)$, that is, the parts of the body DD(a) become connected. The isodensity surface threshold values a_j at which such connections occur are characteristic to the given configuration of the nuclei and to the electronic state. In actual computations, if the density domains are calculated by some *ab initio* technique, then the calculated a_j values and the associated shapes of the isodensity contours $G(a_j)$ are also dependent on the level of quantum chemical approximation, for example, the basis set applied. At a given value a, each density domain $DD_i(a)$ represents a molecular fragment that can be regarded as bound together by chemical bonding at electronic density level a. One may consider each density domain $DD_i(a)$ as the representative of the body of a part of the molecule, *chemically bonded at electronic density level* a. A density domain $DD_i(a)$ that is connected at a higher density value a is likely to be held together by a stronger chemical bonding than one that is connected only at some lower threshold value a. At very high density values a, one finds only disconnected atomic neighborhoods as nearly spherical density domains $DD_i(a)$; at such extreme density levels only individual atomic fragments can be regarded as undivided entities. By contrast, at low density threshold values a, all the nuclei of the entire molecule are enclosed by a single contour surface $G(a)$ of a single density domain $DD_1(a)$; hence, at such an electronic density level a the entire molecule can be regarded as chemically bonded, held together within a single envelope. The density domain concept of chemical bonding is three-dimensional, it is based on the connection between the various parts of the molecular body at various density values. The actual bonding is not modeled by some infinitely thin "bonding channel", such as the conventional model of a formal chemical bond represented by a line. On the contrary, *chemical bonding is attributed to a whole region of the space,* occupied by an electronic charge cloud, where the density is at or above some threshold value a. The density domain bonding concept is *molecular,* it avoids the oversimplifications involved in traditional bonding concepts reduced to formal atom pairs.

The three-dimensional aspect of the DD approach is reminiscent of some features of an earlier approach where chemical bonding is described by Berlin diagrams [261]. Note, however, that in contrast to the formal bonding and antibonding regions of Berlin diagrams [261], which have different signs, the electronic density has the same sign everywhere, and the electronic charge present in any part of a molecular neighborhood can be regarded as bound to the molecule. The DD approach describes the relative contributions of various regions of space to chemical bonding of molecular fragments, by considering the pattern of their stepwise interconnection as the isodensity contour parameter a is varied.

As one of the simplest examples for density domain analysis [262], in Figure 2.2 selected families $\{DD_i(a_j)\}$ of density domains of the water molecule are shown, as calculated using the GAUSSIAN 90 *ab initio* program [253] and the GSHAPE 90 molecular shape analysis program [254], employing a 6-31G** basis set. The density threshold values a represented in the figure are given in atomic units and refer to the 6-31G** level of *ab initio* calculations. These and most alternative Gaussian basis sets have been designed to provide a good description of valence shell properties of molecules, somewhat at the expense of the representation of the core regions near the nuclei. Partly due to this factor, and also due to the lack of taking into account the full electron correlation, relativistic effects, and the proper "cusp" condition at the nuclei (not well represented by Gaussian type AO functions), the wavefunctions and hence the electronic charge densities, are of somewhat lower quality close to the nuclei than in the peripheral regions of the

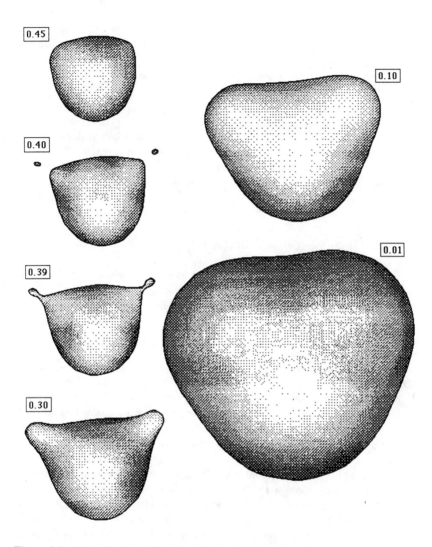

Figure 2.2   Selected families $\{DD_i(a_j)\}$ of density domains of the water molecule, as calculated with the GAUSSIAN 90 *ab initio* program [253] and the GSHAPE 90 molecular shape analysis program [254], using a 6-31G** basis set. There are only two topologically different types of families of density domains: either a single density domain, or a family of three density domains. The sequence of topologically distinct cases provides a topological description of chemical bonding.

molecules. Nevertheless, the main topological patterns and features of the shapes of density domains are well represented.

There are only two topologically different types of families of density domains of water: either a single density domain,

$$\{DD_i(a)\} = \{DD_1(a)\},\qquad\qquad\qquad\qquad\qquad\qquad (2.11)$$

that is, $DD(a)= DD_1(a)$, or a family of three density domains,

$$\{DD_i(a')\} = \{DD_1(a'),\ \ DD_2(a'),\ \ DD_3(a')\}.\qquad\qquad\qquad (2.12)$$

The first family displayed, $\{DD_i(a_1)\}= \{DD_1(a_1)\}$, contains only a single density domain $DD_1(a_1)$ of density threshold value $a_1=0.45$ at the given 6-31G** level of *ab initio* calculations. The corresponding body $DD(a_1)$ has only one component, it is a topological ball, and the only nucleus it encloses is that of the oxygen atom. At a lower density threshold value a_2, separate density domains about the hydrogen nuclei appear, hence a topologically different family

$$\{DD_i(a_2)\}= \{DD_1(a_2),\ \ DD_2(a_2),\ \ DD_3(a_2)\}\qquad\qquad\qquad (2.13)$$

of density domains is obtained, as illustrated by the case of contour of density $a_2=0.40$. This topological type of three disjoint topological balls persists only in a rather narrow density range, and at density $a_3=0.39$ one finds again a family

$$\{DD_i(a_3)\} = \{DD_1(a_3)\}\qquad\qquad\qquad\qquad\qquad\qquad (2.14)$$

that contains only a single density domain $DD_1(a_3)$, a topological ball. This ball, as well as all other bodies $DD(a)$ with a contour density value less than 0.39, encloses all three nuclei of the water molecule. One may say that at and below the density value of 0.39 the water molecule is bound together, whereas at the density value of $a=0.40$ the water molecule is disconnected. The essential features of the full, 3D bonding pattern of a water molecule can be characterized by the density domains of a sequence of three families of topological objects: the first family is a set of a single ball, the second family is a set of three balls, and the third family, again, is a set of a single ball. Of course, the first and the third families are topologically equivalent: by a continuous deformation one can transform the molecular body $DD(a)$ at density threshold $a=0.45$ to the ones at and below the density threshold of $a=0.39$, without cutting or gluing. (Here topological equivalence refers to the intuitively most obvious, ordinary metric topology of the three-dimensional space, that, in our case, distinguishes connected and disconnected objects. Note that alternative choices for topologies will be described in the forthcoming chapters which are suitable for discriminating between minute details in the shapes of the density domains. Based on such topologies, the first and third cases may well be topologically nonequivalent.)

It is worth emphasizing that there are only *finitely many* (actually only three) *density threshold ranges* of the water molecule which are distinguishable using the simplest topological criterion of identifying maximum connected components, the density domains $DD_i(a)$ of the molecular bodies $DD(a)$.

It is advantageous to assign the nuclear labels A, B, etc., enclosed within each density domain to the domain $DD_i(a_j)$, and to use a more descriptive notation

$DD_i(a_j, A, B, \ldots)$. In terms of this notation, the density domain description of the bonding of the water molecule can be given as

$$\{DD_1(a_1,O)\}; \quad \{DD_1(a_2,H), DD_2(a_2,O), DD_3(a_2,H)\}; \quad \{DD_1(a_3,H,O,H)\} \tag{2.15}$$

Of course, there are important shape variations within each of the above three density threshold ranges. For example, the "catface" shape at $a=0.30$ changes to a much rounder shape at $a=0.01$, and at much lower density thresholds the shape becomes essentially spherical. All these shape changes can be described in detail by finer topological techniques, where some geometrical criteria, such as curvature variations, are used to define patches and domains on each MIDCO. These patches and domains on the MIDCO surfaces are used to generate topologies, which can distinguish fine details of shape within the framework of the Shape Group Method (SGM), discussed in later chapters.

2.5 Functional Groups and their Shapes as Quantum Chemical Concepts: A Density Domain Criterion for Functional Groups

The density domain approach provides a quantum chemical criterion for deciding the identity of functional groups. This approach also serves as a tool for the direct shape analysis of functional groups. Before elaborating on these aspects of the DD approach, we shall consider an illustrative example: the density domain shape analysis of chemical bonding in the ethanol molecule, CH_3CH_2OH, as calculated with a 6-31G* basis set [263], using the GAUSSIAN 90 [253] *ab initio* and the GSHAPE 90 [254] molecular shape analysis programs. Some of the results are shown in Figure 2.3 (high density thresholds) and in Figure 2.4 (low density thresholds). The single nucleus density domain $DD_1(a_1,O)$ of the oxygen atom that appears first at a high density threshold and the next $DD(a)$ body of the methyl carbon and oxygen atoms $\{DD_1(a_2,C), DD_2(a_2,O)\}$ are not shown in the figures.

The first $DD(a)$ pattern shown in Figure 2.3 is the family

$$\{DD_1(a_3,C), DD_2(a_3,C), DD_3(a_3,O)\} \tag{2.16}$$

of the three density domains of the heavy nuclei, followed by families showing the sequential appearance of the hydrogen density domains. At the $a_j=0.407$ threshold all nuclear neighborhoods are present as density domains, resulting in the family

$$\{DD_1(a_j,H), DD_2(a_j,H), DD_3(a_j,H), DD_4(a_j,H), DD_5(a_j,H),$$
$$DD_6(a_j,H), DD_7(a_j,C), DD_8(a_j,C), DD_9(a_j,O)\}. \tag{2.17}$$

The first merger of density domains in the above sequence occurs at the threshold value of $a_j=0.390$, where the density domain of the OH proton merges with that of the oxygen atom. This results in the family

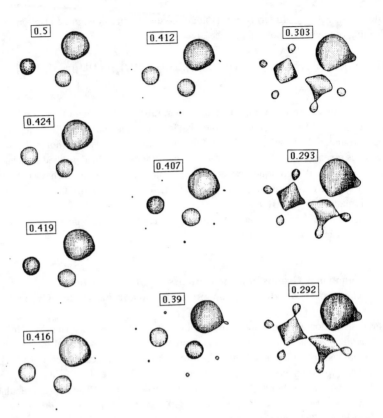

Figure 2.3 Some of the high density threshold density domains of the ethanol molecule, CH_3CH_2OH, as calculated with a 6-31G* basis set, using the GAUSSIAN 90 [253] *ab initio* and the GSHAPE 90 [254] molecular shape analysis programs.

$$\{DD_1(a_j,H), \quad DD_2(a_j,H), \quad DD_3(a_j,H), \quad DD_4(a_j,H),$$
$$DD_5(a_j,H), \quad DD_6(a_j,C), \quad DD_7(a_j,C), \quad DD_8(a_j,O,H)\} \tag{2.18}$$

of density domains representing the formal molecular body $DD(a_j)$ at this density threshold. This is the first example in the given sequence for the formation of a density domain of a typical functional group, the OH group. This density domain $DD_8(0.390,O,H)\}$ provides a density-based justification for regarding the OH functional group as a chemical entity of separate identity: there exists a range of density threshold values at which the corresponding two nuclei, those of O and H, are surrounded by MIDCO's, separating O and H from all other nuclei of the molecule. This very property is used for the identification and a more detailed characterization of the functional groups of chemistry [109,262].

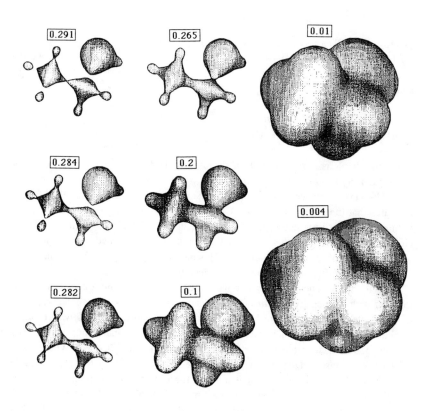

Figure 2.4 Low density threshold density domains of the ethanol molecule, CH_3CH_2OH, as calculated with a 6-31G* basis set, using the GAUSSIAN 90 [253] *ab initio* and the GSHAPE 90 [254] molecular shape analysis programs.

It is interesting to note that in the process of gradually decreasing the threshold a of density domains, it is this most acidic proton of the ethanol molecule that loses its separate density domain first, at the highest density threshold value where a merger of density domains occurs. In other words, the minimal electronic charge density is the highest between this proton and another nucleus of the molecule, in spite of the fact that as the most acidic proton of the molecule, this is the proton most easily donated to proton acceptors. The explanation of this apparently counterintuitive observation lies in the high electronegativity and the resulting, more extensive high density domain about the oxygen nucleus. What we observe is in fact the high electron density about the oxygen which engulfs this particular proton.

The next two topologically different families of density domains, represented by the threshold values of $a_j=0.303$ and $a_{j'}=0.293$, correspond to the two-step merger of methylenic carbon and the two methylenic hydrogen density domains. The resulting two families,

$$\{DD_1(a_j,H),\ DD_2(a_j,H),\ DD_3(a_j,H),\ DD_4(a_j,H),$$
$$DD_5(a_j,C),\ DD_6(a_j,C,H),\ DD_7(a_j,O,H)\}, \qquad\qquad (2.19)$$

and

$$\{DD_1(a_{j'},H),\ DD_2(a_{j'},H),\ DD_3(a_{j'},H),\ DD_4(a_{j'},C),$$
$$DD_5(a_{j'},C,H,H),\ DD_6(a_{j'},O,H)\} \qquad\qquad (2.20)$$

of density domains represent the formal molecular bodies $DD(a_j)$ and $DD(a_{j'})$ for the two density threshold values.

The very existence of the density domain $DD_5(a_{j'},C,H,H)$ in the last family indicates that the methylene group, CH_2, is a chemical entity of separate identity within the ethanol molecule, if one uses the density domains as criterion. There clearly exists a MIDCO that separates the carbon and the two hydrogen nuclei from all other nuclei of the ethanol molecule. This justifies regarding the methylene group as a functional group within the ethanol molecule.

In the most general sense, a collection of all nuclei within a density domain, together with the density domain $DD_i(a_j)$ can be regarded as a functional group of the molecule at the density threshold a_j. This provides a physically motivated choice [109,262] within the general scheme [264] for a functional group analysis of various molecular arrangements in the nuclear configuration space M. Whereas not all density domains $DD_i(a_j)$ and their enclosed nuclei are likely to be transferable from molecule to molecule without altering the topology of the density domain, nonetheless, there is justification for this approach. Having a MIDCO separating the given group of nuclei from the rest of the nuclei of the molecule does indicate a stronger chemical linkage among the local charge densities surrounding these given nuclei than the linkage between this group and the rest of the molecule. This is a property characteristic of the more common functional groups of chemistry. This concept of functional group does not necessarily coincide with the concept conventionally used by chemists. In order to avoid confusion, we shall indicate that we deal with density domain (DD) functional groups whenever ambiguity may arise. In practical computations, the quality of the calculated charge density (e.g., 6-31G*) should also be indicated, especially for the high density threshold ranges.

In our example of the ethanol molecule, the next topological change is the merger of the methyl carbon density domain with one of the methyl hydrogen density domains, resulting in the density domain family

$$\{DD_1(a_j,H),\ DD_2(a_j,H),\ DD_3(a_j,C,H),\ DD_4(a_j,C,H,H),\ DD_5(a_j,O,H)\} \qquad (2.21)$$

as shown by the $DD(a_j)$ body at $a_j=0.292$.

It is of some interest that according to the given 6-31G* level of quantum chemical calculations, the methyl group *does not appear* as a formal functional group of separate identity within this molecule. It is evident from the DD(a) body at threshold $a=0.291$ that the next merger occurs between the $DD_3(a_j,C,H)$ and $DD_4(a_j,C,H,H)$ domains of the previous density threshold, resulting in the new family of density domains

$$\{DD_1(a,H), DD_2(a,H), DD_3(a,C,H,C,H,H), DD_4(a,O,H)\}, \qquad (2.22)$$

as shown in Figure 2.4. This observation implies that in the ethanol molecule no density domain of the type $DD_i(a,C,H,H,H)$ exists. That is, there exists no MIDCO that separates the carbon and three hydrogen nuclei of the formal methyl group from all other nuclei of the ethanol molecule. If one uses the natural electronic density domains as criterion, *the methyl group has no separate identity within the ethanol molecule.* As a density domain functional group, a methyl group is not a part of the ethanol molecule.

Anyone with nostalgia for tradition may obtain some consolation from our next observation: the ethyl group *is* a density domain functional group of the ethanol molecule. This is evident if one follows the changes in the density domains as the density threshold a is further lowered. One by one, the two remaining hydrogen density domains are linked up with the earlier $DD_3(a,C,H,C,H,H)$ density domain of $a=0.291$, as shown by the molecular bodies $DD(a)$ at $a=0.284$ and at $a=0.282$. The DD family making up the molecular body $DD(0.282)$ consists of only two density domains,

$$\{DD_1(0.282,C,H,H,H,C,H,H), DD_2(0.282,O,H)\}. \qquad (2.23)$$

Evidently, $DD_1(0.282,C,H,H,H,C,H,H)$ is a density domain of the ethyl functional group, whereas $DD_2(0.282,O,H)$ is a density domain of the hydroxyl functional group, both manifesting a separate identity at this density threshold value. Both the ethyl and the hydroxyl groups are present as DD functional groups of the ethanol molecule, a reassuring observation in view of the conventional name of the molecule, *ethyl alcohol.*

The next important topological change is the merger of these two density domains, as illustrated by the example with a threshold value of $a=0.265$.   The family of density domains has only a single member,

$$\{DD_1(0.265,C,H,H,H,C,H,H,O,H)\}. \qquad (2.24)$$

The *bonding sequence* as given by the topological sequence of families of density domains is now complete, and this sequence provides a detailed, three-dimensional description of bonding in the ethanol molecule.

At this stage, the nuclei of the molecule are enclosed by a single density contour, a feature present at all lower density thresholds. Although no fragmentation of density domains can occur at lower thresholds, nevertheless, further important shape changes occur at threshold values below $a=0.265$, from the shape of a puppy dog visiting a lamppost at $a=0.200$ to the shape of an irregular potato at $a=0.004$, leading to an eventual spherical shape at very low densities.

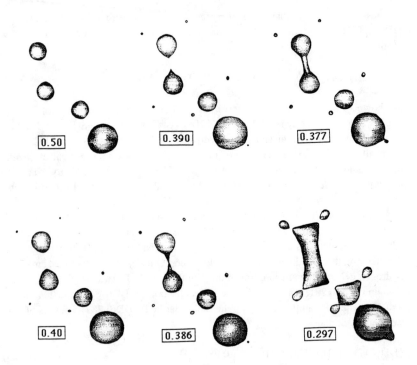

Figure 2.5 Some of the high density threshold density domains of the most stable conformation of allyl alcohol, $CH_2=CH-CH_2-OH$, as calculated with the GAUSSIAN 90 and GSHAPE 90 programs, using a 6-31G* basis set.

Although these shapes are all different, they all represent the same topological molecular body of a single density domain,

$$DD(a) = DD_1(a). \qquad (2.25)$$

The shape changes of this single density domain can be described in detail using the Shape Group Method (SGM) [155-158], discussed in later chapters of this book.

Our last example of DD analysis of bonding is that of the most stable conformation of allyl alcohol, $CH_2=CH-CH_2-OH$, as calculated with the GAUSSIAN 90 [253] and GSHAPE 90 programs [254], using a 6-31G* basis set.

The allyl alcohol molecule shows many of the typical local shape features of organic molecules, and it represents a useful test case for topological shape analysis methods [262,263]. A sequence of topologically different families of density domains of $CH_2=CH-CH_2-OH$ is shown in Figures 2.5 and 2.6. This set of DD's shows the potential of the approach for a detailed analysis of molecular bonding patterns, exhibiting many chemically interesting features. Starting with a high

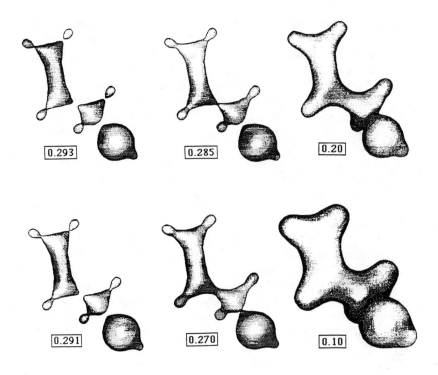

Figure 2.6 Some of the low density threshold density domains of the most stable conformation of allyl alcohol, $CH_2=CH-CH_2-OH$, as calculated with the GAUSSIAN 90 and GSHAPE 90 programs, using a 6-31G* basis set.

density threshold, and considering a gradual decrease of the isodensity contour value a, the appearance of the DD's of the heavy nuclei is followed by those of the hydrogen nuclei. One should note that the hydrogenic DD that appears last (at a=0.390) is that of the OH proton. Interestingly, this is the hydrogenic density domain that loses its separate identity the earliest (at a=0.377), and other hydrogenic DD's join those of neighboring carbon atoms at much lower electron density threshold values (the first one at a=0.297). The electronegative oxygen lowers the electron density about the proton of the OH group, consequently, this hydrogenic density domain appears last. On the other hand, the oxygen also increases the electron density in the space between the two nuclei, consequently, the first (i.e., highest threshold) density domain linking a proton with another nucleus also involves the OH group. Note, however, that as expected, the DD's of the double bonded pair of carbons join first (at the highest density threshold where a merger occurs, at a=0.386).

Some of the more common functional groups of organic chemistry have density domains which persist over a wide range of density threshold values. As it has also been found in the case of ethanol, by far the most persistent   DD of the allyl alcohol is that of the OH group. This density domain preserves its separate identity within a very wide range of density thresholds, from a=0.377 to a value slightly below a=0.285 (as the threshold a is decreased). The CH_2=CH vinyl group has its separate DD in the approximate range of density thresholds between a=0.290 and a=0.286, whereas the CH_2 methylene group has its own DD in the range between thresholds a=0.292 and a=0.286. The DD of the CH_2=CH-CH_2- allyl group is also prominent as a separate entity within the threshold range of a=0.285 and a=0.271. At a threshold value of a=0.270, the allyl alcohol molecule is bound together as a *single body* DD(a), by having the density domains of the allyl and hydroxyl groups joined at this threshold. At and below the density threshold value of a=0.270, the molecule has only a single density domain and the formal molecular body DD(a) is a topological sphere. The topological sequence of DD families provides a detailed description of bonding within the allyl alcohol molecule. The example of allyl alcohol will be used again in later chapters of the book for illustrating the application of the Shape Group Methods (SGM) to the study of details of shape and shape variations.

The density domain analysis of several molecules has indicated that the appearance of individual DD's of hydrogen nuclei and their joining with the DD of a neighboring atom or functional group follow a trend, especially, if electronegative heteroatoms are involved. This trend, called the "Last-First Rule", can be stated as follows.

The Last-First Rule:
If the threshold density is gradually decreased in a DD analysis of a given molecule, then the order of joining of various hydrogenic DD's to neighboring DD's tends to be the reversed order of the appearance of these hydrogenic DD's.

In particular, the hydrogenic DD that appears last is likely to join a neighboring DD first, an observation that justifies the name of the rule. As has been discussed above, this is the typical case, illustrated and explained by the example of allyl alcohol, where the neighboring DD is that of an electronegative oxygen. This trend, however, is not strictly followed in all instances. For example, in the case of the ethanol molecule, the last few hydrogenic DD's appear within a narrow density range, and the DD of the OH hydrogen does not appear last, but is the first joining a neighboring heavy atom DD. Hence, in a strict sense, the rule is violated for the calculated example of ethanol. Nevertheless, the DD of the OH hydrogen joins a neighboring DD much before this happens to the hydrogenic DD that appeared first, hence the trend, in an approximate sense, is a useful guideline. Similar, but less reliable Last-First Rules describe the trends for other, nonhydrogenic atomic neighborhood DD's. For example, for carbon atom DD's the order of their appearance and the order of their joining to neighboring DD's are expected to be approximate reverses of each other; however, more deviations from the rule are likely for nonhydrogenic atomic neighborhood DD's.

The more important functional groups have separate DD identity within an

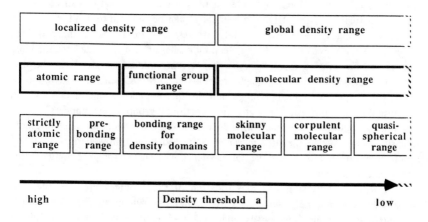

Figure 2.7 Classification of density domains according to ranges of the density threshold a.

intermediate range of a values. However, some general trends can be observed in the entire range of density threshold values. In all molecules, the bonding pattern shows some characteristic density ranges where typical shape features occur. The classification of DD's according to various ranges of the density threshold parameter a is illustrated in Figure 2.7. In the high density range only individual nuclear neighborhoods appear as disconnected DD's (i.e., there is precisely one nucleus within each density domain which appears). This density range can be referred to as the *atomic range*. In a lower density range some nuclear neighborhoods join to form DD's containing two or more nuclei, but not all nuclei of the molecule fall yet within a common density domain. In this density range one finds the various functional groups as individual entities, hence this range can be referred to as the *functional group range*. This is the range where the bonding pattern of density domains is revealed, consequently, the alternative term *bonding range for density domains* also gives a valid characterization for this range. In a lower density range all nuclei of the molecule fall within a common DD (i.e., the essential molecular pattern of bonding is established), and this range is called the *molecular density range*. The atomic and the functional group ranges together form the *localized range,* and when juxtaposed to this range, the molecular density range can be regarded as a *global density range.*

The molecular density range can be subdivided into subranges. At the highest threshold values within the molecular range, the DD has some local "neck" region, that is, there is at least one topological belt or some other multiply connected set on the surface of the density domain along which the surface is not locally convex. (Note that within any multiply connected set there are loops which cannot be contracted to a single point.) Within this density subrange the molecule appears as a single entity, but the corresponding body is "skinny", hence this range can be referred to as the *skinny molecular range.* At even lower densities no such neck regions occur, but for most molecules there is at least one local nonconvex region along the surface of the DD. This subrange can be referred to as the *corpulent molecular range.* At very low (possibly infinitesimally low) densities the MIDCO's of all molecules, even those of long chain molecules are convex; the corresponding range can be referred to as the *quasi-spherical molecular range.* For the allyl alcohol molecule all five ranges appear, although no MIDCO of the last two ranges (i.e., the corpulent molecular range and the quasi-spherical range) is shown in Figure 2.6. In contrast, for water no functional group range appears due to symmetry: both protonic neighborhoods join the oxygen DD at the same threshold density value, hence the atomic range is immediately followed by the skinny molecular range. For water, the lowest density MIDCO shown belongs to the corpulent molecular range.

Further subdivision and classification of these ranges are possible. For example, the atomic range can be subdivided into two subranges, one where all the atomic density domains are convex, and another where at least one DD is no longer convex, just before the joining of neighboring density domains occurs. The first subrange is the *strictly atomic range,* whereas the second one can be referred to as the *prebonding range.*

Based on the theory of catastrophes as applied to the electronic density by Collard and Hall [265], considerable effort has been made by Bader and co-workers [266-270] to use the gradient of the electron density to reduce the quantum chemical concept of bonding to the simple, conventional picture of chemists: bonds as lines interconnecting atoms in a molecule. This approach has appeal in its simplicity and in its promise to justify the symbolism of bonds drawn as lines by generations of chemists. However, this bonding line model does not coincide with a description of bonding based on the actual energy criteria of bonding [271] and as shown by Cioslowski and co-workers, in some molecular cases it predicts bonds where no actual bonding is present [272-277]. If our goal is the description of the shape of the actual, fuzzy body of the electronic density, then a pattern of lines as formal bonds between nuclei is insufficient for this purpose. By contrast, the density domain approach provides a description of chemical bonding in terms of the actual immersion of the nuclear arrangement within an electronic cloud and the broad interfacing of molecular fragments bound to one another.

CHAPTER

3

APPLIED TOPOLOGY:
THE MATHEMATICS OF THE ESSENTIAL

By using an analogy of human characters, topology is the more tolerant, broad-minded, more intuitive sibling of geometry. Topology bends where geometry breaks. Tolerance, however, does not mean imprecision or sloppiness: topology provides a rigorous description of qualitative aspects of objects. Topology offers elegant shortcuts to results which would take painstaking effort or could not be reached using geometrical tools. The term "qualitative" should not be taken as a sign of weakness or limitation. Most of our accumulated scientific information, even numerical data, are based on qualitative understanding and qualitative knowledge. In fact, direct human observation by our senses almost always results in a qualitative comparison and exact coincidence of an observed property with some reference seldom occurs. In most physical and chemical measurements the value we obtain and consider as a *quantitative numerical* result is, in fact, almost always a *qualitative* answer indicating that with some probability the measured property can be classified qualitatively as belonging to a numerical interval. This numerical interval and the associated probability are defined by the error bar which is only *centered* on a numerical value. In geometrical terms, the central numerical value corresponds to a point, whereas the probabilistic interval can be regarded as a topological domain. By choosing finer and finer qualitative classification criteria (e.g., narrower and narrower intervals), one obtains a more and more unaccommodating, "crisper" description and, in the limit, one obtains a fully quantitative, rigid, geometrical description. Few objects of nature measure up to the exact specifications of geometry, and the more tolerant topology can provide a better, more accommodating, and more faithful representation of real objects. When compared to

49

geometrical models, topological models have fewer inherent constraints that make them different from the objects they represent, hence, topological models have a better potential to describe reality. In particular, the nonrigidity of molecules and the quantum mechanical Heisenberg uncertainty relation suggest topology as an eminently suitable tool for the description of molecular shape.

The analysis of the three-dimensional topology of formal molecular bodies has not yet become a standard tool of chemistry, in contrast to the widely used tools of graph theory of formal chemical bonds. However, the perception of topology as a difficult and somewhat esoteric subject is rapidly changing among chemists. The exceptional versatility of applied topological methods is becoming more and more apparent and many useful chemical applications of topology have been proposed ([59-72, 103-112], and references therein).

From the chemist's point of view, an important aspect of topology is its ability to provide a discrete description of some essential properties of continuous problems. This may be used for extracting the most relevant chemical information from complicated molecular models in a form especially advantageous for computer analysis. Algebraic topology is particularly suitable for this task: the essential properties of continuous functions are described by algebraic means (e.g., by groups) and various other topological invariants. The topological shape analysis methods are applicable to formal molecular surfaces, for example, to molecular isodensity contours (MIDCO's), to molecular electrostatic potential (MEP) contour surfaces (MEPCO's), or to Van der Waals surfaces (VDWS'), leading to a family of *algebraic shape groups*. The resulting discrete description, a *numerical shape code*, can be stored, processed, and compared by a computer, a possibility that offers an algorithmic approach to such problems as the nonvisual analysis of 3D molecular shapes, computer evaluation of molecular shape similarity, and the study of shape complementarity of reacting biomolecules.

3.1   Topology as "Rubber Geometry"

Topology is one of the most powerful modern chapters of mathematics. As a mathematical discipline, topology provides a solid foundation and link for many diverse areas of mathematics, from continuous functions, through set theory, to algebra and geometry. Topology is an ideal tool for performing a rigorous analysis of qualitative features, for extracting the essential information, and to explore various levels of similarity among objects. Topology provides efficient methods and computational techniques for a systematic and mathematically rigorous *disregard* of those features that are *not relevant* to the problem at hand. This is especially important if one is interested in the underlying essential features of a complicated object like a molecule, while being less interested in its unimportant details, or in details that are only artifacts of the model used. Topology is often compared to geometry: its colloquial label, "rubber geometry", does express some of the essence of topology. For an object made from rubber, just as for any abstract topological object, many geometrical changes are possible without changing the identity of the object. In such objects, the exact distance between two points (a geometrical concept) is unimportant, whereas the important properties are connectedness, continuity, and neighborhoods (which are topological concepts). Topology can be regarded as the mathematics of the essential: the topological properties of a physical object are often those which are preserved as long as the identity of the object is preserved.

In this book we shall be concerned only with a very limited selection of the elements of topology, relevant to the basics of topological shape analysis of molecules. All the tools we shall use will be described in sufficient detail in the book. However, for readers interested in more details of the fundamentals, some introductory and advanced texts are listed among the references [113-122].

One of the fundamental tools of topology is the concept of continuous deformation: deforming an object without cutting or gluing its parts. Such a continuous deformation is called a *homotopy*. Two objects which are related to each other by a homotopy are *homotopically equivalent*. A homotopy is allowed to bring two points into a single point during the deformation, for example, a homotopy may contract a disk or a ball into a single point. This property can lead to the loss of important shape information, for example, a homotopy may convert an entire object into a single point. If, however, we require that different points stay different during the transformation, and that the assignment of the initial set of points to the final set of points, as well as the reversed assignment, are continuous, then many essential features are necessarily preserved. A transformation that can accomplish this is called a *homeomorphism*. In other words, a homeomorphism is a "reversible" continuous transformation that converts each point of the original object to a unique point of the new object, while the same is true for the inverse transformation, that is, for the reversed process of continuously transforming the new object back to the old one. Notice that this condition excludes the possibility of cutting or gluing, since continuity of the forward and reverse transformations does not allow points that are very (infinitesimally) near to each other to become very distant, or very distant points to be placed very near to each other. If there is a homeomorphism between two objects, we then say that the two objects are *topologically equivalent*.

One can find some insight into the topological approach by considering the standard example used by topologists: the case of the doughnut and the coffee cup. As illustrated in Figure 3.1, if the doughnut is made from an easily deformable material, a continuous deformation can convert it into a coffee cup. Note, however, that no homeomorphism exists between the doughnut and an ordinary potato: neither punching a hole in the potato nor cutting the doughnut through to its central hole is a continuous transformation. The doughnut and the potato are not topologically equivalent, however, the doughnut and the coffee cup are related by a homeomorphism, consequently, in the ordinary topological sense, they are topologically equivalent. Hence the saying: a topologist is someone who cannot distinguish a doughnut from a coffee cup. This might appear discouraging for the reader: what is the power of a mathematical treatment for shape analysis that cannot distinguish the shapes of two such objects of obviously very different forms? Fortunately, topology can provide as detailed a shape characterization as desired, where the doughnut and the coffee cup are recognized as having very different shapes indeed. For this purpose, the principal tool of topology is a redefinition of

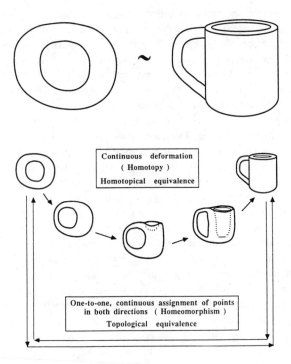

Figure 3.1   The doughnut and the coffee cup: an example for continuous deformation (homotopy) and topological equivalence (homeomorphism). If the doughnut is made from an easily deformable material such as soft clay, then a continuous deformation can convert it into a coffee cup. Such a continuous deformation is called a homotopy, implying that the doughnut and the coffee cup are homotopically equivalent. A homotopy may unite several points, for example, it may contract a disk into a single point; however, special homotopies may also preserve the distinctness of each point of the object. The deformation can be chosen so that it assigns to each point of the doughnut a unique point in the coffee cup and *vice versa*. Furthermore, the assignment of points is continuous in both directions: in the deformation and also in the reverse deformation turning the coffee cup back to the doughnut. A transformation with such properties is called a homeomorphism, that is the very condition for topological equivalence. The doughnut and the coffee cup are related by a homeomorphism, that is, they are topologically equivalent.

continuity. By adopting various levels of "graininess" a transformation that is continuous on one level may be discontinuous on another level. That is, two objects that are topologically equivalent on one level can be topologically nonequivalent on a different level. Hence, by adopting a suitable choice for continuity, fine details of shape can be monitored and distinguished by topological means.

In fact, general topology takes an even bolder step. By explicitly declaring which sets of points can participate in transformations that are going to be called continuous, these sets are said to form a topology, and from then on continuity is regarded within this framework. However, for this general scheme to make practical sense (to retain some resemblance to ordinary continuity and not to be self-contradictory), the chosen sets must fulfill some conditions. These conditions are described in the next section.

3.2   Some Basic Concepts of Topology

One of the most difficult aspects of mathematics is not what it describes but what it ignores. Mathematical analysis requires definitions which specify some properties and imply some others, but the abstract objects so defined have no additional properties. Here lies the difficulty: when compared to real objects, the abstract mathematical objects we need for precise statements have, in fact, very few properties. Our imagination, trained on ordinary objects, often brings up images of objects which in some of their properties match those defined, but these actual objects usually have many more properties. It is often difficult to avoid being influenced by these additional properties and to keep our intuition and reasoning restricted to the properties present in the abstract object. The temptation to draw conclusions from actual real life analogies is particularly strong when applying mathematical models for molecular shape analysis. When dealing with complex objects of nature, the simplistic approach that everything is important can no longer be used, and the task is to ignore the unimportant detail while retaining the essential in a mathematically rigorous way. One avenue to such abstraction is provided by topology. When referring to the topological definitions and when using them, the reader is advised to keep in mind that *nothing more is meant than said.*

Notations. It is advantageous to use the concise and precise mathematical notations of set theory. A general set will be denoted by X, its elements by $\mathbf{x}, \mathbf{y}$, etc., the fact that $\mathbf{x}$ is an element of X is denoted by $\mathbf{x} \in X$, and if $\mathbf{x}$ is not an element of set X, we can state this as $\mathbf{x} \notin X$. In topology, the concept of space is rather general and, depending on the context, ordinary sets such as the collection of all points of a potato may be referred to as spaces. Instead of the expression "element $\mathbf{x}$ of set X" we shall often use the term "point $\mathbf{x}$ of space X". We shall use the following common notations: $A \cap B$ for intersection ("overlap") of two sets A and B, $A \cup B$ for the union of A and B (considering the members of A and B as belonging to a common family), and $A \subset B$ for stating that A is a subset (subfamily) of B.

Metric and metric space.   The concept of metric may be regarded as the generalization of the familiar concept of distance. If a real-valued function $d(\mathbf{x}, \mathbf{y})$ is defined for every pair of elements, $\mathbf{x}, \mathbf{y} \in X$, of a set X, and if this function d has the properties

(i) $d(\mathbf{x}, \mathbf{y}) \geq 0$ for any pair $\mathbf{x}, \mathbf{y} \in X$, $\hspace{2cm}$ (3.1)

(ii) $d(\mathbf{x}, \mathbf{y}) = 0$ if and only if $\mathbf{x} = \mathbf{y}$, $\hspace{2.2cm}$ (3.2)

(iii) $d(\mathbf{x}, \mathbf{y}) = d(\mathbf{y}, \mathbf{x}),$ (3.3)

(iv) $d(\mathbf{x}, \mathbf{z}) \leq d(\mathbf{x}, \mathbf{y}) + d(\mathbf{y}, \mathbf{z}),$ (3.4)

for any three elements $\mathbf{x}, \mathbf{y}, \mathbf{z} \in X$, then this function $d(\mathbf{x}, \mathbf{y})$ is a *metric* of the set X. The above relations are the natural properties of distance: never negative, zero if and only if the two points coincide, symmetric (the distance of **x** from **y** is the same as the distance of **y** from **x**), and fulfill the (3.4) triangle inequality (the sum of the lengths of two sides of a triangle cannot be less than the length of the third side).

A nonempty set X, provided with such a   *metric* $d(\mathbf{x}, \mathbf{y})$ for every pair of its elements, $\mathbf{x}, \mathbf{y} \in X$, is called a   *metric space.*

Open sets,   closed sets,   relative complement,   and   continuity of functions. In a metric space Y, a subset A of Y is regarded as an *open set* if around every point of A there exists some (perhaps very tiny) ball that is still within set A. Informally, an open set does not contain its boundary points (e.g., an open potato is the potato without its skin where the skin is thought to be infinitely thin). The *relative complement* of A in space Y is the set A^c of all points of Y which are not in A. The relative complement can be written as $A^c = Y \backslash A$. A subset C of set Y is a *closed set* in Y if the relative complement C^c of C is an open subset in Y. The *closure* clos(A) of a set A is the smallest closed set that contains A.

A *function* f can be regarded as an assignment of points **x** of a set X to points **y** of a set Y, expressed by the notation

f: $X \rightarrow Y.$ (3.5)

Note that the above notation refers to the simultaneous assignment of all points of set X to some points of set Y. For an individual point pair **x** and **y** one may write $f(\mathbf{x}) = \mathbf{y}$.

A function f may assign points **x** of a set X to some points **x'** of the *same* set X, a fact expressed as f: $X \rightarrow X.$ For the special case of the *identity function* I,

I: $X \rightarrow X,$ (3.6)

where $I(\mathbf{x}) = \mathbf{x'}$, the points **x** and **x'** are the same (i.e., $I(\mathbf{x}) = \mathbf{x}$).

Informally, the image of a set A according to the function f is the set generated from A by f, whereas the inverse image of a set B in Y is the set needed in X to get B when applying f. In more precise terms, the *image* $B = f(A)$ of a subset A of X is the set of all points $\mathbf{y} = f(\mathbf{x})$ of Y for which $\mathbf{x} \in A$. The *inverse image* $f^{-1}(B)$ of a subset B of Y is the set $A = f^{-1}(B)$ of all points $\mathbf{x} \in X$ for which $f(\mathbf{x}) \in B$; in colloquial terms, the inverse image of B is the set A where B comes from.

Informally, a function f: X → Y is *continuous* if, for any point pair **x, x'** that are near to each other in X, their images **y**=f(**x**) and **y'**=f(**x'**) are also "near enough" to each other in Y. A precise definition of continuity can be given as follows: a function f: X → Y is *continuous* if and only if the inverse image of every open subset B of Y is an open subset A of X.

Above we have given definitions of open sets and continuity only within a metric space. For the definition of open sets we have needed the concept of distance (defined in a metric space) in order to be able to specify the radius of some balls around each point. However, distance is not a topological concept and we cannot rely on it in a general topological treatment. To get around this, we can exploit the fact that open sets have certain fundamental properties and these very properties may be used to *define* which sets are to be regarded as open sets. We shall review these properties below. If we select a family **T** of sets which can be declared open sets without conflict with these general properties, then this family **T** of subsets of X is said to form a *topology* on X. Then continuous functions can also be defined by the same formal condition as given above (the inverse image of any open set is open), but now open sets do not have to obey any conditions based on distance, in fact, we can do away with distance altogether.

Topology and topological space. Topology is the most general mathematical theory of open sets. Within a set X a *topology* **T** is given if one specifies in a consistent manner which sets are to be regarded as open sets. Below we give a formal definition of a topology in terms of the three fundamental properties of open sets (also valid, of course, for open sets within a metric space):
    A class **T** of subsets of X,

$$\mathbf{T} = \{T_\alpha : T_\alpha \subset X\}, \tag{3.7}$$

is a *topology* on X, if

  (i) $X, \varnothing \in \mathbf{T}$ $\hspace{3cm}$ (3.8)

    where $\varnothing$ is the empty set (a family with no members),

  (ii) $\underset{\beta}{\cup} T_\beta \in \mathbf{T}$ $\hspace{3cm}$ (3.9)

    for any number of sets in **T**, and

(iii) $T_\alpha \cap T_\beta \in \mathbf{T}$ $\hspace{3cm}$ (3.10)

    for any two sets $T_\alpha, T_\beta \in \mathbf{T}$.

If a set X is provided with a topology **T**, then the pair (X,**T**) is called a *topological space*. Since there are many ways of selecting a topology **T**, from a given set X we can generate many topological spaces. We can no longer simply say that a set is open or closed: we have to specify which topology is considered. Elements T_α of topology T are called T-*open sets*. A set C is T-*closed* if its complement $C^c = X \backslash C$ is a T-open set. A set is regarded as an open or a closed set depending on the topology, in fact, if T_1 and T_2 are two different topologies on set X, then a set $A \subset X$ may be T_1-open but T_2-closed. The simple expressions "open set" and "closed set" can only be used if the choice of the topology is clear from the context.

On a given set X some of the topologies are interrelated in a hierarchical way. Consider two topologies T_1 and T_2 on a set X, where every T_1-open subset of X is also T_2-open. Then the topology T_1 is a subfamily of T_2 (i.e., $T_1 \subset T_2$) and we say that topology T_1 is *weaker (coarser)* than T_2, or alternatively, we say that T_2 is *stronger (finer)* than T_1. Two topologies are *not comparable* if neither is weaker than the other.

A set $N \subset X$ is called a *neighborhood* of point $x \in X$ if and only if there exists an open set $G \in T$ such that $x \in G \subset N$.

A subfamily $B \subset T$ is called a *base* for topology T if and only if every open set $G \in T$ is a union of some sets in B.

A family S of sets is a *subbase* for a topology T if and only if finite intersections of members of S (the "overlaps" of various collections of a finite number of sets from subbase S) form a base B for topology T. Of course, each member of a subbase S is a T-open set of the topology T of base B.

A *metric topological space* (X, T) is the set X provided with a *metric topology* T, where the T-open sets are precisely those which are open in the ordinary sense according to some metric d, defined on X.

Now we can define *continuous functions* in topological terms. For two topological spaces (X_1, T_1) and (X_2, T_2) a function φ from set X_1 to set X_2 is *continuous* if and only if the inverse image of every T_2-open subset of X_2 is a T_1-open subset of X_1:

$$\text{for every } G \in T_2 , \; \varphi^{-1} (G) \in T_1. \tag{3.11}$$

A function φ is *one-to-one* if it assigns a unique element $\varphi(x) = y \in X_2$ to each element $x \in X_1$. A function φ is *onto* if every element $y \in X_2$ is assigned to some element $x \in X_1$. A function φ is *bijective* if it is both one-to-one and onto.

A function φ is a *homeomorphism* if it is bijective and both φ and φ^{-1} are continuous,

$$\varphi, \; \varphi^{-1} \in C, \tag{3.12}$$

that is, if both φ and φ^{-1} are elements of the class C of continuous functions on X_1 and X_2, respectively.

Two topological spaces (X_1, T_1) and (X_2, T_2) are *topologically equivalent* or *homeomorphic* if there exists a function $f: X_1 \to X_2$ which is bijective and both f and f^{-1} are continuous.

Among a family $\{a,b,c,d,...\}$ of objects, some objects may be regarded as equivalent with respect to some of their properties. Such an equivalence can be expressed by a relation, called an *equivalence relation*. An equivalence relation, here denoted by the symbol $\sim$, must have three properties:

reflexive, $a \sim a$,

symmetric, if $a \sim b$ then $b \sim a$, and

transitive, if $a \sim b$ and $b \sim c$, then $a \sim c$.

In a family of objects an equivalence relation generates a classification: we can collect those objects which are equivalent to one another into *equivalence classes.* Each object a belongs to precisely one equivalence class.

A property is said to be *topological* or a *topological invariant* if it is a property of all topological spaces in an equivalence class generated by the equivalence relation "topologically equivalent".

A topological space (X, T) is *disconnected* if X is a union of two, nonempty, disjoint T-open subsets,

$$X = A \cup B, \quad A, B \neq \varnothing, \quad A \cap B = \varnothing, \quad A, B \in T. \tag{3.13}$$

A topological space (X, T) is said to be *connected* if it is not disconnected. A connected open subset is also called a *domain,* although the word "domain" is often used in a more general sense.

Informally, a subset A of X is said to be *contractible to a point* if a continuous contraction of the set A to a single point is possible without requiring any point of A to pass through points not in set X.

A *topological sphere* is an object that is topologically equivalent to a sphere.

An n-dimensional set X is *simply connected* if and only if every k-dimensional $(k \leq n)$ topological sphere ^kS in X is contractible to a point.

Compactness is a generalization of properties of closed and bounded intervals, applicable to higher dimensional sets. It is defined in terms of various *coverings* of sets.

Consider a subset $A \subset X$. If family $F = \{F_i\}$ is a class of open subsets of set X such that

$$A \subset \bigcup_i F_i \tag{3.14}$$

then F is called an *open cover* of set A. If family F contains only a finite number of F_i subsets, then F is called a *finite cover.*

A subset A of a topological space X is *compact* if every open cover of A contains a finite subcover. For example, a sphere is compact; but if we remove a point from the sphere, the generated *punctured sphere* is no longer compact. Missing a single point from a sphere may not appear all that important, but due to the resulting lack of compactness, it has important topological consequences.

Elements of algebraic topology, chains, cycles, and homology groups. The techniques of algebraic topology are applicable for the description and concise characterization of molecular shapes. Homology theory is usually introduced in the context of polyhedra, providing a tool for the analysis of the interrelations among various faces, edges, and vertices of these polyhedra. However, the methods of homology theory are not restricted to polyhedra, since they are applicable to objects of curved surfaces, such as formal molecular bodies. Homology groups are suitable to describe the topological relations among the elements of various subdivisions of surfaces. These subdivisions can be very general, such as those obtained by irregular patches, their boundary lines and the joining points of these lines on various surfaces, including formal molecular surfaces, MIDCO's, and VDWS'. One should notice that in a topological setting the object under study does not have to have planar boundaries in order to be topologically equivalent to a polyhedron. Hence, we shall be able to use the terminology of polyhedra for a topological analysis of the shapes of general irregularly shaped surfaces.

In a general n-dimensional polyhedron, a *p-dimensional face* is called a *p-face*. For example, in a regular tetrahedron embedded in the ordinary three-dimensional space there are four 2-faces (ordinary faces), six 1-faces (edges), and four 0-faces (vertices). The simplest faces are *simplexes*, for example, the simplest two-dimensional face a polyhedron can have is a triangle, that is a two-dimensional simplex. If homology theory is directly introduced on objects that are not polyhedra, one may use a *cellular decomposition* of the object where the *p-dimensional cells* or, in short, the *p-cells* are analogous to the p-dimensional faces of a polyhedron, and the development of homology theory is analogous to that in terms of polyhedra and their faces. An *open cell* of dimension p is homeomorphic to the open unit ball of a p-dimensional Euclidean space.

First we shall take a systematic look at polyhedra in terms of simplexes. A *geometrical p-simplex* $S(p,i)$ of any positive dimension $(p > 0)$ is a set of points $\mathbf{x}_\alpha^{(i)}$

$$S(p,i) = \{\mathbf{x}_\alpha^{(i)}\} \tag{3.15}$$

defined in terms of $p+1$ points $\mathbf{p}_0^{(i)}, \mathbf{p}_1^{(i)} \dots \mathbf{p}_p^{(i)}$ of the n-dimensional Euclidean space ^nE as a *convex combination*:

$$\mathbf{x}_\alpha^{(i)} = \sum_{k=0}^{p} t_{\alpha k}\, \mathbf{p}_k^{(i)}, \tag{3.16}$$

where

$$\sum_{k=0}^{p} t_{\alpha k} = 1 \tag{3.17}$$

and

$$0 < t_{\alpha k} < 1, \qquad (k = 0, 1, \dots p), \tag{3.18}$$

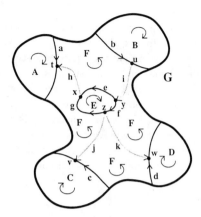

Figure 3.2   An example for the subdivision of the surface of a 3D object into 2-cells (surface patches), 1-cells (line segments on the boundaries of the patches) and 0-cells (joining points of the line segments). The patches themselves can be defined by physical conditions (for example, by ranges of values of some local molecular property such as electrostatic potential), or by some geometrical condition, such as the ranges of local curvatures of the surface. In the latter case, an interesting combination of geometrical and topological treatments is obtained: a geometrical condition on the local curvature is the basis for a topological characterization of shape by homology groups.

and where the set $\{(\mathbf{p}_1^{(i)} - \mathbf{p}_0^{(i)}),\ (\mathbf{p}_2^{(i)} - \mathbf{p}_0^{(i)}),\ ...\ (\mathbf{p}_p^{(i)} - \mathbf{p}_0^{(i)})\}$ of vectors is linearly independent (i.e., not all p + 1 points fall on the same (p-1)-dimensional hyperplane).

The zero-dimensional case is special: a geometric 0-simplex $S(0,i)$ is a single point by definition.

A *geometric simplicial p-complex*   k(p)   is a finite set of disjoint q-simplices of the n-dimensional Euclidean space ^nE, where q = 0, 1, ... p, such that, if

$$S(q,i) \in k(p) \qquad\qquad (3.19)$$

then all faces of $S(q,i)$ are in k(p), and there are no two distinct simplexes in k(p) which have all their faces the same.

The *polyhedron* $|k(p)|$ is the point set union of all points of all simplexes $S(q,i)$ which belong to the simplicial complex   k(p).

Of course, for a general polyhedron not all 2-faces must be 2D simplexes (i.e., triangles). For example, all 2-faces of a cube are squares. Furthermore, for a more general object of curved surface, the 2-faces (ordinary faces), 1-faces (edges) and 0-faces (vertices) of a polyhedron are replaced by the 2-cells (surface patches), 1-cells (line segments on the boundaries of the patches), and 0-cells (joining points of the line segments) of the object, respectively.

In Figure 3.2 a formal molecular surface G and its subdivision into various cells are shown. The surface G   is subdivided into five locally convex domains, denoted by A, B, C, D, and E, and the remaining sixth domain F, where the local curvature properties of the surface can be characterized as being of the "saddle type". Note that here we use the term "domain" in a colloquial and somewhat imprecise sense. The "domain" F is multiply connected since not every loop can be contracted within F into a single point. For example, such noncontractible loops in F can be found around domain E. Hence, F is not a true domain in the strict mathematical sense; nevertheless, we shall use this colloquial terminology for F. However, the division lines h, i, j, and k convert F into a simply connected domain. The 2-cells of the resulting subdivision are A, B, C, D, E, and F; the 1-cells are the line segments a, b, c, d, e, f, g, h, i, j, and k; whereas the 0-cells are the points t, u, v, w, x, y, and z. We shall consider the following index assignments:

$$
\begin{array}{lll}
C(2,1) = A, & C(1,1) = a, & C(0,1) = t, \\
C(2,2) = B, & C(1,2) = b, & C(0,2) = u, \\
C(2,3) = C, & C(1,3) = c, & C(0,3) = v, \\
C(2,4) = D, & C(1,4) = d, & C(0,4) = w, \\
C(2,5) = E, & C(1,5) = e, & C(0,5) = x, \\
C(2,6) = F, & C(1,6) = f, & C(0,6) = y, \\
& C(1,7) = g, & C(0,7) = z, \\
& C(1,8) = h, & \\
& C(1,9) = i, & \\
& C(1,10) = j, & \\
& C(1,11) = k. &
\end{array}
\tag{3.20}
$$

Each cell may be given an *orientation*, formally denoted by +1 or -1. These orientations can be represented by arrows; for example, a curved arrow can be assigned to each 2-cell. An arrow with a counterclockwise rotation when viewed from the interior of the surface can be associated with a +1 orientation, and one with a clockwise rotation can be associated with a -1 orientation. A 1-cell can be assigned an arrow along the corresponding line segment, and one can use an arbitrary (but fixed) convention in assigning orientation numbers +1 or -1 to each arrow. A 0-cell (a point) is usually assigned a formal +1 orientation. For example, in Figure 3.2, the orientations of 2-cells A, B, and E are +1 (counterclockwise when viewed from the interior of G), whereas the orientations of 2-cells C, D, and F are -1 (clockwise when viewed from the interior of G).

If cell C'(p) is the same as cell C(p), except its orientation is reversed, then we may write

$$
C'(p) = - C(p).
\tag{3.21}
$$

In addition to signs, numerical values can also be assigned to cells, and these values can be regarded as coefficients of the cells. Now we are in the position to add and subtract cells and to define formal linear combinations of various cells. The generation of homology groups is based on such linear combinations, called p-*chains*. Whereas these chains are geometrical or topological objects, we may think of these linear combinations as formal "shopping lists" of cells.

A p-dimensional *chain* of p-cells is defined as the following abstract linear combination of $C(p,i)$ cells of a common dimension p:

$$c^p = \sum_{i=1} u_i \, C(p,i) \tag{3.22}$$

where the u_i coefficients are scalars. For example, the 2-chain

$$c^2 = 15 \, C(2,1) + 0 \, C(2,2) + 2 \, C(2,3) - 5 \, C(2,4) + \ldots ,$$

that is,

$$c^2 = 15 \, A + 0 \, B + 2 \, C - 5 \, D + \ldots , \tag{3.23}$$

of the example in Figure 3.2, is a formal linear combination of 2-dimensional cells $C(2,i)$. In general, we may think of a p-chain as a linear combination of *all* 2-cells, some of which may have zero coefficients, such as the 2-cell $B = C(2,2)$ in the 2-chain c^2 of Equation (3.23). It is common to omit cells of zero coefficients when writing these linear combinations. The c^p chains of a common dimension can be added up and multiplied by scalar numbers, resulting in chains of the same dimension.

Within *integer homology theory,* sufficiently versatile for our purposes of molecular shape analysis, we consider only integers as coefficients and multiplying factors of cells and chains.

The family C^p of all such c^p chains of cells,

$$C^p = \{c_a^p\} , \tag{3.24}$$

forms an Abelian group with respect to the addition defined as

$$c_1^p + c_2^p = \sum_{i=1} (u_i + u_i') \, C(p,i), \tag{3.25}$$

where u_i and u_i' are the coefficients of cell $C(p,i)$ in the chains c_1^p and c_2^p, respectively. The neutral element of this group (denoted by 0^p, or simply by 0) is the p-chain that has all of its coefficients equal to zero.

The $\Delta C(p,i)$ *boundary* of a p-dimensional cell $C(p,i)$ is the (p-1)-chain

$$\Delta C(p,i) = \sum_{j} n_{ij}(p-1) C(p-1,j), \tag{3.26}$$

where $n_{ij}(p-1)$ is the *incidence number* between cells $C(p,i)$ and $C(p-1,j)$.

The incidence number $n_{ij}(p-1)$ is zero if the $(p-1)$-cell $C(p-1,j)$ is *not* on the set theoretical boundary of p-cell $C(p,i)$,

$$n_{ij}(p-1) = 0, \qquad\qquad\qquad (3.27)$$

otherwise, *if $C(p-1,j)$ and $C(p,i)$ meet only on one occasion,* then

$$n_{ij}(p-1) = 1 \qquad\qquad\qquad (3.28)$$

if the orientations of the two cells match, and

$$n_{ij}(p-1) = -1 \qquad\qquad\qquad (3.29)$$

if the orientations do not match. That is, *if $C(p-1,j)$ and $C(p,i)$ meet only on one occasion,* then

$$n_{ij}(p-1) = \begin{cases} 0 & \text{if } C(p-1,j) \text{ is not on } \Delta_S C(p,i) \\ +1 & \text{if } C(p-1,j) \text{ is on } \Delta_S C(p,i) \text{ and orientations match} \\ -1 & \text{if } C(p-1,j) \text{ is on } \Delta_S C(p,i) \text{ and orientations don't match,} \end{cases} \qquad (3.30)$$

where $\Delta_S C(p,i)$ denotes the set theoretical boundary of $C(p,i)$, the collection of points falling on the boundary of $C(p,i)$. Note that the set theoretical boundary is a set of points, whereas the boundary defined in Equation (3.26) is a chain.

If cells $C(p-1,j)$ and $C(p,i)$ meet on more than one occasion, then the numbers given by (3.30) give only the *local incidence number* for one occasion; for the k-th occasion the local incidence number is denoted by $n_{ijk}(p-1)$. In this case, the +1 and -1 numbers on the right-hand side of Equation (3.30) must be summed up for each occasion $C(p-1,j)$ and $C(p,i)$ meet, and the sum gives the value of the incidence number:

$$n_{ij}(p-1) = \sum_{k} n_{ijk}(p-1). \qquad\qquad\qquad (3.31)$$

For example, in Figure 3.2, the 2-cell $F = C(2,6)$ stretches over the far side of the surface G, hence it meets the 1-cell $h = C(1,8)$ on two occasions. When F meets h from the upper right-hand side, then there is a mismatch of orientations; hence we obtain a local incidence number of $n_{6,8,1}(1) = -1$. However, when F meets h from the lower left-hand side, then orientations match; hence we obtain a local incidence number of $n_{6,8,2}(1) = +1$. According to Equation (3.31), the sum of these two values, $-1 + 1 = 0$, is the incidence number $n_{6,8}(1)$ for these two cells, $F = C(2,6)$, and $h = C(1,8)$:

$$n_{6,8}(1) = n_{6,8,1}(1) + n_{6,8,2}(1) = -1 + 1 = 0. \qquad\qquad (3.32)$$

In more general, "folded" cases, incidence numbers of higher absolute values can occur, however, these cases are irrelevant to the molecular shape analysis techniques of this book.

The local incidence numbers between 1-cells and 0-cells conform with the following convention: -1 or +1 correspond to the oriented 1-cell leaving or entering the 0-cell, respectively.

In colloquial terms, the boundary $\Delta C(p,i)$ of a set $C(p,i)$ is a chain of $C(p-1,j)$ sets where the coefficients of the linear combination are the incidence numbers between set $C(p,i)$ and sets of dimension one less, $p-1$. For example, the boundary $\Delta E = \Delta C(2,5)$ of 2-cell $E = C(2,5)$ of Figure 3.2 is the 1-chain

$$c^1 = \Delta E = \Delta C(2,5) = \sum_j n_{5,j}(1)C(1,j) =$$

$$= -C(1,5) + C(1,6) + C(1,7) = -e + f + g. \tag{3.33}$$

In the above sum, all other 1-cells participate with a zero coefficient, hence they are omitted.

The boundary of a zero-dimensional cell $C(0,i)$ is zero by definition. The *boundary* of a p-chain $c^p = \sum u_i C(p,i)$ of cells is defined as the (p-1)-chain

$$\Delta c^p = \sum_{i=1} u_i \Delta C(p,i) = \sum_{i,j} u_i n_{ij}(p-1)C(p-1,j), \tag{3.34}$$

i.e., as the linear combination of the boundaries of the cells making up the chain. Since, in the above sum, a given cell $C(p-1,j)$ may enter on several occasions as a contributor to different (p-1)-chains with positive or negative coefficients, these contributions may cancel out and then the overall coefficient of $C(p-1,j)$ may turn out to be zero. In fact, in special cases the coefficients of all cells in the linear combination may turn out to be zero, then we obtain the zero (p-1)-chain 0^{p-1} that may be written simply as 0.

An important property of boundaries is expressed by the relation

$$\Delta\Delta c^p = 0, \tag{3.35}$$

i.e., *the boundary of a boundary is zero*. For example, according to Equation (3.33), the boundary of the boundary of 2-cell E is

$$\Delta\Delta E = \Delta(-e + f + g) = -\Delta e + \Delta f + \Delta g =$$

$$= -(-y + x) + (-y + z) + (-z + x) = 0x + 0y + 0z = 0 \tag{3.36}$$

A *p-cycle* is a p-chain which has no boundary, that is, whose boundary is zero, $\Delta c^p = 0$. For example, in Figure 3.2 the 1-chain $(-e + f + g)$ is a cycle; according to Equation (3.36), the boundary of this chain is zero. The set of all p-cycles forms a subgroup Z^p of group C^p.

The boundary of any (p+1)-chain is a p-cycle but not every p-cycle is a boundary of a (p+1)-chain. A p-cycle is called a *bounding* p-cycle if it bounds something, that is, if it can be given in the form

$$c^p = \Delta \, c^{p+1} \, . \tag{3.37}$$

The bounding cycles are special chains, which bound some other chains on the object (but being cycles, themselves have no boundaries). For example, the cycle $(- e + f + g)$ of Figure 3.2 bounds cell E, hence $(- e + f + g)$ is a bounding cycle. The set of all bounding p-cycles is a subgroup B^p of Z^p.

If for two p-chains c_1^p and c_2^p the *difference* $c_1^p - c_2^p$ is a bounding p-cycle then c_1^p and c_2^p are *homologous*,

$$c_1^p \sim c_2^p, \tag{3.38}$$

where, in the present context, the symbol $\sim$ is used for homology equivalence. Note that if $c_1^p - c_2^p$ is a bounding cycle, then $c_2^p - c_1^p = - (c_1^p - c_2^p)$ is also a bounding cycle (of opposite orientation), hence it is immaterial which chain is written first in the above difference.

If c_1^p is a bounding cycle, and if we take $c_2^p = 0$, the zero cycle, then the difference $c_1^p - c_2^p = c_1^p$ is, evidently, a bounding cycle. By defining $c^p \sim 0$ for every bounding cycle, homology $\sim$ is an equivalence relation within set C^p of p-chains. For the equivalence classes, *homology classes* $[c^p]$, addition is defined as

$$[c_1^p] + [c_2^p] = [c_1^p + c_2^p], \tag{3.39}$$

in particular

$$[c_1^p] + [0] = [c_1^p], \tag{3.40}$$

and

$$[c_1^p] + [- c_1^p] = [0] \, , \tag{3.41}$$

where we simply write $[0]$ for class $[0^p]$.

With the above addition as the group operation, the family of all homology classes of dimension p forms a *group*, the pth (integral) *homology group* H^p of the object on which the various cells are defined. The p-th homology group is the difference group ("quotient" group)

$$H^p = Z^p - B^p. \tag{3.42}$$

For each dimension p the homology groups H^p, just as groups C^p, Z^p, and B^p, are finitely generated Abelian groups. The homology groups are topological invariants.

The *Betti numbers* are important topological invariants which can be used for shape characterization. The p-th *Betti number* b_p is the rank of homology group H^p. For topological objects we encounter in this book, the Betti numbers b_p can be calculated from the following simple relations:

$$b_p = m_p - r_p - r_{p-1} \quad (p>0), \tag{3.43}$$

$$b_o = m_o - r_o. \tag{3.44}$$

Here, m_p is the number of cells $C(p,i)$ of dimension p. Number r_p is the rank of the *incidence matrix* $\mathbb{N}(p)$ of elements

$$\mathbb{N}(1)_{ij} = n_{ij}(p) \tag{3.45}$$

if $p < n$,   otherwise $r_p = 0$.
    For closed n-dimensional surfaces a useful property of Betti numbers is given by the Poincaré *index theorem,*

$$b_p = b_{n-p}. \tag{3.46}$$

The Betti numbers b_p are related to the Euler-Poincaré characteristic χ ,

$$\chi = \sum_{p=0}^{n} (-1)^p \, b_p. \tag{3.47}$$

    The Euler-Poincaré characteristic χ is an important topological invariant of the object.
    The number of holes of a truncated surface G is an easily visualizable topological feature. If the number of holes of G is one or more, then it is equal to $b_1 + 1$.

    A sample calculation of homology groups and Betti numbers.   As an example, here we shall consider the special case of two-dimensional surfaces, such as the example shown in Figure 3.2. First we summarize the special aspects of the two-dimensional case of homology groups in the context of this example. Consider a surface G, for example, a MIDCO G(a), or a truncated surface derived from it, that is subdivided into a family of two-dimensional (2D) domains {C(2,i)}. We assume that each domain C(2,i) is simply connected, that is, every closed loop within C(2,i) can be contracted into a point. This subdivision also generates two additional families, a family {C(1,i)} of the 1D boundary line segments of the C(2,i) domains, and a family {C(1,i)} of 0D points where these line segments meet.
    Each of these C(p,i) sets is given a formal orientation, denoted by +1 or -1. For example, each line segment C(1,i) may be regarded as an arrow, whereas to each domain C(2,i) one may assign a clockwise or counterclockwise rotation, as viewed from the inside of G(a). In all the following applications the choice of these orientations is arbitrary, and the final results are not affected by the choice. Each C(0,i) point may be regarded to have the orientation +1.

Two *incidence matrices,* $\mathbb{N}(1)$ and $\mathbb{N}(0)$ are defined for the pair of 2D and 1D families $\{C(2,i)\}$, $\{C(1,i)\}$, and for the pair of 1D and 0D families $\{C(1,i)\}$, $\{C(0,i)\}$, respectively, by their elements $n_{ij}(p-1)$ taken according to the definition (3.31), where the dimension p is either 2 or 1. Pictorially, these incidence matrices can be represented by *incidence graphs* [158], and can also be augmented with numerical information on the size, e.g., the area of each domain $C(2,i)$.

There are three homology groups for such a 2D surface: H^2, H^1, and H^0, and their ranks, the Betti numbers b_2, b_1, and b_0 can be calculated from relations (3.43) and (3.44).

For the cellular subdivision in Figure 3.2 the following incidence matrices are obtained:

	t	u	v	w	x	y	z	
	0	0	0	0	0	0	0	a
	0	0	0	0	0	0	0	b
	0	0	0	0	0	0	0	c
	0	0	0	0	0	0	0	d
	0	0	0	0	1	-1	0	e
$\mathbb{N}(0) =$	0	0	0	0	0	-1	1	f
	0	0	0	0	1	0	-1	g
	1	0	0	0	-1	0	0	h
	0	-1	0	0	0	1	0	i
	0	0	1	0	0	0	-1	j
	0	0	0	1	0	0	-1	k

(3.48)

and

	a	b	c	d	e	f	g	h	i	j	k	
	1	0	0	0	0	0	0	0	0	0	0	A
	0	-1	0	0	0	0	0	0	0	0	0	B
$\mathbb{N}(1) =$	0	0	-1	0	0	0	0	0	0	0	0	C
	0	0	0	1	0	0	0	0	0	0	0	D
	0	0	0	0	-1	1	1	0	0	0	0	E
	1	-1	-1	1	-1	1	1	0	0	0	0	F

(3.49)

Above and on the right-hand side of these matrices we have indicated in bold face letters the short-hand symbols for cells of Figure 3.2 corresponding to the columns and rows of the matrices. From relations (3.43) and (3.44) we can calculate the Betti numbers by the equivalent formula

$$b_p = m_p - r_p - r_{p-1} \quad (0 \le p \le n = 2), \tag{3.50}$$

where m_p is the number of cells of dimension p, and where

$$r_n = r_{-1} = 0, \tag{3.51}$$

otherwise r_p is the rank of incidence matrix $\mathbb{N}(p)$.

The seven columns of $\mathbb{N}(0)$ are not independent since their sum is the 11-dimensional zero vector; however, any six of these columns are linearly independent, hence the rank $r_0 = 6$. Similarly, the first five rows of $\mathbb{N}(1)$ add up to the sixth row vector, but any five of these six row vectors are linearly independent, hence the rank $r_1 = 5$. Since in the example $m_2 = 6$, $m_1 = 11$, and $m_0 = 7$, we obtain

$$b_2 = 6 - 0 - 5 = 1, \tag{3.52}$$

$$b_1 = 11 - 5 - 6 = 0, \tag{3.53}$$

and

$$b_0 = 7 - 6 - 0 = 1 \tag{3.54}$$

for the three Betti numbers of the complete surface G shown in Figure 3.2. The value of the zero-dimensional Betti number, $b_0 = 1$, indicates that we deal with a surface that is a single piece, whereas the value of the two-dimensional Betti number, $b_2 = 1$, indicates that the surface is closed, (i.e., it has no holes). As we shall see later, the most useful shape information can be deduced from the value of the one-dimensional Betti number b_1.

The two-dimensional (n=2) surface G is closed and the Poincaré index theorem, $b_p = b_{n-p}$, holds.

In the many applications of homology theory to molecular shape analysis, we shall need only the various Betti numbers and the Euler-Poincaré characteristic χ, whereas the actual homology groups will not be used directly. Nevertheless, these homology groups can be calculated from the information presented. The Betti numbers b_p are the ranks of the corresponding homology groups H_p, and in our ("torsion-free") case, the knowledge of these numbers is sufficient for the construction of the homology groups. In the example shown in Figure 3.2, the two homology groups H^2 and H^0 of the molecular surface G are both isomorphic to the additive group of integers, wheres the one-dimensional homology group H^1 is isomorphic to the trivial group.

Consider now a modification of surface G, by removing domain E. We may think of this operation as a truncation of a MIDCO $G(a)$, from where a locally convex domain is excised. Note that cell E is an open set (in the usual sense of a metric Euclidean space, that is, E is a T-open set in the context of the metric topology T). Consequently, the 1-cells e, f, and g, as well as the 0-cells x, y, and z are not affected by this truncation. Consequently, the resulting truncated surface has the same incidence matrix $\mathbb{N}(0)$ as the original surface G, but we shall have a new incidence matrix $\mathbb{N}(1)$ between the family of 1-cells and the new family of 2-cells:

	a	b	c	d	e	f	g	h	i	j	k	
	1	0	0	0	0	0	0	0	0	0	0	A
	0	-1	0	0	0	0	0	0	0	0	0	B
$\mathbb{N}(1) =$	0	0	-1	0	0	0	0	0	0	0	0	C
	0	0	0	1	0	0	0	0	0	0	0	D
	1	-1	-1	1	-1	1	1	0	0	0	0	F

$$(3.55)$$

The numbers $m_1 = 11$ and $m_0 = 7$ of the 1D and 0D cells, respectively, have not changed; however, we obtain the new value of $m_2 = 5$ for the number of 2D cells. For the incidence matrix $\mathbb{N}(0)$ the rank is $r_0 = 6$, as before, and the value r_1 has not changed either. The five row vectors of the new incidence matrix $\mathbb{N}(1)$ are linearly independent, hence we obtain that the rank is $r_1 = 5$. By substituting these values into Equation (3.50), we obtain the following Betti numbers:

$$b_2 = 5 - 0 - 5 = 0, \qquad (3.56)$$

$$b_1 = 11 - 5 - 6 = 0, \qquad (3.57)$$

and

$$b_0 = 7 - 6 - 0 = 1. \qquad (3.58)$$

As a result of the removal of domain E, only the two-dimensional Betti number has changed, from 1 to 0, indicating that the new, truncated surface is no longer a closed surface, it has at least one hole. In fact, the value of $b_1 = 0$, in combination with $b_2 = 0$, tells us that there is one hole, which turns out to be in the place of the missing cell E.

For the truncated surface the Poincaré index theorem, $b_p = b_{n-p}$, does not apply. The two homology groups H^2 and H^1 of the truncated surface are both isomorphic to the trivial group, whereas the homology group H^0 is isomorphic to the additive group of integers.

Consider now another truncation of G, typical in MIDCO shape analysis.

Eliminate all five locally convex domains, A, B, C, D, and E. As before, the families of 0-cells and 1-cells, hence the incidence matrix $\mathbb{N}(0)$ are not affected by this truncation of G, i.e., $m_1 = 11$, $m_0 = 7$, and $r_0 = 6$, as before. However, for the new surface the number of 2-cells has changed to $m_2 = 1$, consequently, the new incidence matrix $\mathbb{N}(1)$ is different,

$$\begin{array}{ccccccccccc} a & b & c & d & e & f & g & h & i & j & k \end{array}$$

$$\mathbb{N}(1) = \begin{array}{ccccccccccc} 1 & -1 & -1 & 1 & -1 & 1 & 1 & 0 & 0 & 0 & 0 \end{array} \quad F \qquad (3.59)$$

being a simple row vector of rank $r_1 = 1$.

Using Equation (3.50), we obtain the following Betti numbers:

$$b_2' = 1 - 0 - 1 = 0, \tag{3.60}$$

$$b_1 = 11 - 1 - 6 = 4, \tag{3.61}$$

and

$$b_0 = 7 - 6 - 0 = 1. \tag{3.62}$$

The homology groups H^2 and H^0 of the new truncated surface are isomorphic to the trivial group and the additive group of integers, respectively, whereas the homology group H^1 is isomorphic to the Abelian group of four free generators, denoted by g_1, g_2, g_3, and g_4. The elements of this group H^1 can be written in the form $k_1g_1 + k_2g_2 + k_3g_3 + k_4g_4$, where k_1, k_2, k_3, and k_4 are integers.

The two-dimensional Betti number is zero, $b_2 = 0$, indicating that the new, truncated surface is not a closed surface, it has at least one hole. The one-dimensional Betti number b_1 is four. This value of $b_1 = 4$ indicates that there are $4 + 1 = 5$ holes, in the places of the missing cells A, B, C, D, and E. The zero-dimensional Betti number $b_0 = 1$ indicates that the truncated surface has not fallen apart into pieces, it is still a single piece.

In all the examples discussed above, locally concave or convex domains have been excised from the molecular surface G shown in Figure 3.2. For the actual molecular surface G, none of these truncations has led to a separation of G into pieces. In contrast, a rather different result is obtained if we decide to eliminate domain F together with its internal subdivision lines h, i, j, and k. This step is equivalent to the excision of the locally saddle type region of a MIDCO surface G(a). The remaining, truncated object is no longer a single piece, it is a collection of disconnected surfaces, although one may treat this collection as a single topological object. For this object the 2-cells are A, B, C, D, and E, the 1-cells are a, b, c, d, e, f, and g, whereas the 0-cells are the set t, u, v, w, x, y, and z of the original surface G. For the resulting new topological object one obtains the following incidence matrices:

$$\mathbb{N}(0) = \begin{array}{ccccccc} t & u & v & w & x & y & z \\ 0 & 0 & 0 & 0 & 0 & 0 & 0 \\ 0 & 0 & 0 & 0 & 0 & 0 & 0 \\ 0 & 0 & 0 & 0 & 0 & 0 & 0 \\ 0 & 0 & 0 & 0 & 0 & 0 & 0 \\ 0 & 0 & 0 & 0 & 1 & -1 & 0 \\ 0 & 0 & 0 & 0 & 0 & -1 & 1 \\ 0 & 0 & 0 & 0 & 1 & 0 & -1 \end{array} \begin{array}{c} \\ a \\ b \\ c \\ d \\ e \\ f \\ g \end{array} \qquad (3.63)$$

and

$$\mathbb{N}(1) = \begin{array}{ccccccc} a & b & c & d & e & f & g \\ 1 & 0 & 0 & 0 & 0 & 0 & 0 \\ 0 & -1 & 0 & 0 & 0 & 0 & 0 \\ 0 & 0 & -1 & 0 & 0 & 0 & 0 \\ 0 & 0 & 0 & 1 & 0 & 0 & 0 \\ 0 & 0 & 0 & 0 & -1 & 1 & 1 \end{array} \begin{array}{c} \\ A \\ B \\ C \\ D \\ E \end{array} \qquad (3.64)$$

The new cell numbers are $m_2 = 5$, $m_1 = 7$, and $m_0 = 7$, whereas the ranks of the new incidence matrices $\mathbb{N}(1)$ and $\mathbb{N}(0)$ are $r_1 = 5$, and $r_0 = 2$, respectively. By substituting these values into Equation (3.50), we obtain the following Betti numbers:

$$b_2 = 5 - 0 - 5 = 0, \qquad (3.65)$$

$$b_1 = 7 - 5 - 2 = 0, \qquad (3.66)$$

and

$$b_0 = 7 - 2 - 0 = 5. \qquad (3.67)$$

The zero-dimensional Betti number $b_0 = 5$ indicates that we have five separate pieces, none of which is a closed surface ($b_2 = 0$), and the number of holes is the minimum ($b_1 = 0$) compatible with the above information, i.e., each piece has one hole. This hole should be interpreted as follows: if any one of these pieces is curled up at the edges, then one can view this piece as a ball that has a hole.

Two of the homology groups, H^2 and H^1, of the new disconnected object are both isomorphic to the trivial group. The zero-dimensional homology group H^0 is isomorphic to the Abelian group of five free generators. If these generators are denoted by g_1, g_2, g_3, g_4, and g_5, then the elements of homology group H^0 can be written in the form $k_1 g_1 + k_2 g_2 + k_3 g_3 + k_4 g_4 + k_5 g_5$, where $k_1, k_2, k_3, k_4,$ and k_5 are integers.

   Convexity and curvature properties.   In the above discussion and examples we have already used the concepts of convexity and locally convex domains in an intuitive manner. Whereas our goal is to provide a topological shape characterization for molecules, we shall often use geometrical tools at intermediate steps toward a topological description. These steps often involve the concepts of convexity, curvature, and a characterization of critical points of functions.

   A function $f(x)$ is *convex* (convex from below) over an interval $[x_1, x_2]$ if and only if

$$f(\alpha_1 x_1 + \alpha_2 x_2) \leq \alpha_1 f(x_1) + \alpha_2 f(x_2) \tag{3.68}$$

for any choice of coefficients α_1 and α_2 fulfilling the conditions

$$\alpha_1 + \alpha_2 = 1, \tag{3.69}$$

and

$$\alpha_1, \alpha_2 \geq 0. \tag{3.70}$$

A function $f(x)$ is *concave* (concave from below) if in condition (3.68) the strict inequality of reversed direction holds for coefficients α_1 and α_2 fulfilling (3.69) and (3.70).

   A set A is *convex* if and only if for any two elements x_1, x_2,

$$x_1, x_2 \in A, \tag{3.71}$$

and for any pair of coefficients α_1 and α_2 fulfilling conditions (3.69) and (3.70) the relation

$$\alpha_1 x_1 + \alpha_2 x_2 \in A \tag{3.72}$$

holds. That is, set A is convex if and only if for any point pair $x_1, x_2 \in A$ any point of the $[x_1, x_2]$ line segment is also an element of set A.

   A point x, given as the linear combination

$$x = \alpha_0 x_0 + \alpha_1 x_1 + \ldots \alpha_k x_k \tag{3.73}$$

is a *convex combination* of points

$$x_0, x_1, \ldots x_k \in A \tag{3.74}$$

if and only if the α_i scalars fulfill the following conditions:

$$\alpha_0 + \alpha_1 + ... + \alpha_k = 1, \tag{3.75}$$

and

$$\alpha_i \geq 0, \quad \text{for every } i = 0, 1 ... k. \tag{3.76}$$

The set B of all such convex combinations, with reference to a fixed set of points (3.74),

$$B = \{x = \alpha_0 x_0 + \alpha_1 x_1 + ... + \alpha_k x_k\} \tag{3.77}$$

is a *closed* k-*dimensional simplex* if and only if the set

$$\{(x_1 - x_0), (x_2 - x_0), ... (x_k - x_0)\} \tag{3.78}$$

of vectors is linearly independent. If condition (3.76) is replaced by the strict inequality

$$\alpha_i > 0 \quad \text{for every } i = 0, 1, ... k, \tag{3.79}$$

then set B is an *open* k-*dimensional simplex* .

Critical points of functions. If we think of the independent variable(s) of a function as being plotted along a horizontal line, or a horizontal plane, (or a horizontal hyperplane, if there are more than two such variables), and think of the function values as being plotted vertically, then, informally, a critical point is a point where the function under consideration is locally horizontal. In more precise terms, a *critical point* of a twice continuously differentiable function $E(r)$ is a point r_C where the gradient of function $E(r)$, defined for points r of a set X, vanishes,

$$g(r_C) = 0 . \tag{3.80}$$

The *Hessian matrix* $\mathbb{H}(r_C)$ of a function $E(r)$ is the matrix of the second partial derivatives of the function, defined by matrix elements

$$H_{ij}(r) = \partial^2 E(r)/\partial r_i \partial r_j \tag{3.81}$$

(usually defined only at critical points, that is, where the first derivatives are zero). The eigenvalues of the Hessian matrix characterize the critical point r_C. If the rank of the Hessian matrix $\mathbb{H}(r_C)$ is smaller than the dimension n of X, then r_C is called a *degenerate critical point.*

The number λ of negative eigenvalues of matrix $\mathbb{H}(r_C)$ is called the *index* of the critical point r_C.

The Hessian matrix and the index of the critical point are tools for a precise characterization of curvature properties, important in shape analysis.

3.3   Some Relevant Aspects of Knot Theory

The topologist's concept of a knot is different from the knots used in everyday practice. Consider a rope with your favorite knot in it. Unless the two ends of the rope are joined, the knot, however tight, can be untied by allowing one of its ends to move along the rope. Hence, the initial arrangement is topologically and "knot theoretically" equivalent to the untied rope stretched along a straight line, with no knot in it. Consequently, a topologist does not regard a single, open-ended rope as a knot, no matter how tightly your favorite knot is tied in this rope. On the other hand, if the two ends of the rope are joined, then in most occasions one cannot untie the knot without cutting the rope: in such a case no continuous deformation of the rope can convert it into a circle. In fact, there are infinitely many different ways of having knots in a rope with endpoints joined, and without cutting the rope, these knots cannot be converted into one another.

In a topological sense, *a knot is a homeomorphic image of a circle in the* 3D *space.* Consequently, in a strict sense, all knots are topologically equivalent. However, this equivalence does not mean that all knots can be deformed into one another without passing through themselves (i.e., without cutting and gluing). Knots may differ in their "knottedness", the way they are embedded in the three-dimensional space. Some of the simplest knots are the *trefoil knots* and the *figure eight knot,* shown in Figure 3.3. Most but not all knots are chiral: no motion of the rope can convert a chiral knot into its mirror image. The simplest chiral knots are the left-handed and right-handed trefoil knots, shown in Figure 3.3.

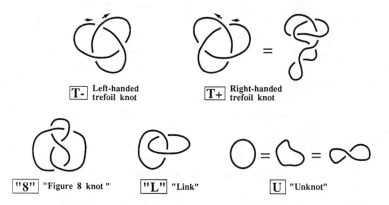

Figure 3.3   The simplest chiral knots, the left-handed and right-handed trefoil knots T_- and T_+ : no motion of the rope can convert a chiral knot into its mirror image. An orientation is specified along the rope of the two trefoil knots. Also shown are the topologically achiral figure eight knot "8", the simple link L, and the unknot U.

| Left-handed crossing X- | Avoided crossing Xo | Right-handed crossing X+ |

Figure 3.4   A convention for the characterization of the handedness of crossings is shown. An orientation, indicated by an arrow, can be assigned to the rope in an arbitrary manner, both choices lead to the same final result. A crossing of the rope is *left-handed* or *right-handed* if the arrows show a pattern of the crossing of the thumb and index finger of the left or right hand, respectively. Note that the classification of the crossing does not change if the orientation is reversed, since the change of orientation affects both arrows on the two rope segments crossed.

For the characterization of the handedness of these knots, one can use the convention shown in Figure 3.4. An orientation can be assigned to the rope in an arbitrary manner; either of the two choices leads to the same final result. The orientation is indicated by an arrow. A crossing of the rope is *left-handed* if the arrows show a pattern of the crossing of the thumb and index finger of the left hand, and the analogous rule applies for a *right-handed crossing*. Note that the classification of the crossing does not change if the orientation is reversed, since this change affects both arrows on the two rope segments crossed.

We place the rope in such a way that no two crossings appear on top of each other. The projection of such an arrangement to a plane is called a *regular projection*. The minimum number of crossings one can have for a regular projection of a given knot is called the *crossing number* of the knot. The crossing number of both trefoil knots is 3. In such a projection, shown in the top left diagram of Figure 3.3, all crossings of the left-handed trefoil knot are left handed, whereas for the analogous projection of the right-handed trefoil knot all crossings are right handed. This property is not general: for some knots both left- and right-handed crossings can occur, even in regular projections. Note that the top right diagram of Figure 3.3 is a knot-theoretically equivalent representation of the right-handed trefoil knot where there are five crossings: two of these crossings can be easily removed by flipping a segment of the rope.

The figure eight knot shown in a geometrically chiral representation is in fact a topologically achiral knot: by appropriate movement of the rope one can obtain the mirror image of the arrangement shown. In fact, a geometrically chiral arrangement can be converted into its mirror image without ever becoming geometrically achiral. These observations demonstrate that *geometrical chirality* and *topological chirality* are fundamentally different, in fact, there are many different levels of topological chirality [71]. Note, however, that the figure eight knot can be arranged in a geometrically achiral manner, but this requires an arrangement with more crossings than the crossing number of 4; for example, a geometrically achiral arrangement

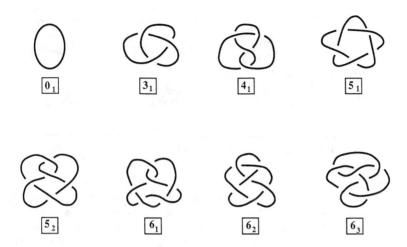

Figure 3.5   Regular projections and the usual knot theoretical symbols of knots of crossing numbers less than seven. For each pair of topologically chiral knots only one topological enantiomer is shown. The geometrical properties of the projections can be misleading. For example, the "figure eight knot", denoted by 4_1, appears chiral since the actual 3D geometrical arrangement of the string is chiral. However, by moving the left-hand side loop of the string over the rest of the knot, followed by minor shifts of the string, one can obtain the mirror image of the arrangement shown. Consequently, the figure eight knot is not topologically chiral.

of S_4 symmetry has eight crossings [71].

In the strict sense neither the link L of Figure 3.3 nor the simple loop U, informally called the "unknot", is a knot, nevertheless, the generic term "knot" is often used for them. Links are formed by more than one rope, which ropes may or may not be knotted, whereas the unknot U is not knotted.

In Figure 3.5, regular projections of knots with crossing numbers less than seven are shown, together with their symbolic notations commonly used in knot theory. For each topologically chiral knot only one of the two topological enantiomers is shown.

There are various knot invariants which are independent of the arrangement of the rope of a given knot. Some of these invariants are polynomials determined from the crossings of the projection of the knot. Note that for the calculation of these invariants the knot does not have to be arranged in any special way beyond the requirement of having a regular projection. In particular, there is no need to generate a projection where the number of crossings is minimum (i.e., equal to the actual crossing number of the knot). The construction of such a polynomial is based on information (the crossing pattern) that does depend on the actual arrangement and projection of the knot, nevertheless, as long as the projections are regular, the resulting polynomial is invariant to changes in the arrangement. Whereas a rearrangement of the rope can eliminate some crossings and can introduce some new ones, surprisingly, the resulting polynomial is unaffected by these changes. Such a

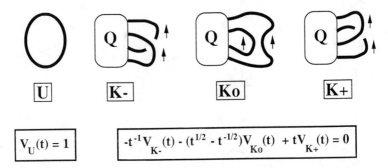

Figure 3.6   The recursive rules for the construction of the Jones polynomials are shown. By definition, the Jones polynomial $V_U(t)$ of the unknot U is 1. Consider three knots, denoted by K_-, K_0, and K_+, which are identical under the cover Q and differ only in their exposed parts, having a left-handed crossing, an avoided crossing, and a right-handed crossing, respectively. Their Jones polynomials, $V_{K_-}(t)$, $V_{K_0}(t)$, and $V_{K_+}(t)$ are interrelated by the equation shown. These rules are sufficient to construct the Jones polynomial for any finite knot. (See comment on normalization convention in the text.) The polynomial can be derived from the crossing information of any placement of the given knot; interestingly, the same polynomial is obtained for every placement, even for highly folded and twisted arrangements of the string, with projections showing a large number of crossings. The Jones polynomial is a knot invariant.

polynomial is an *invariant of the knot.*

One of the most interesting and useful knot polynomials is the Jones polynomial $V_K(t)$ discovered in 1985 [244]. This is a polynomial of a rather general type: it can contain both positive and negative fractional powers of the variable t. The Jones polynomial $V_K(t)$ has the following intriguing property: the polynomial $V_K(t)$ of a knot K and the polynomial $V_{K^\Diamond}(t)$ of the $K^\Diamond$ mirror image of the knot K are related in a simple way:

$$V_{K^\Diamond}(t) = V_K(t^{-1}). \qquad (3.82)$$

The above rule serves as a powerful tool for testing chirality of knots: if the replacement of variable t by its reciprocal t^{-1} in the polynomial $V_K(t)$ changes the polynomial into a different one, that is, if

$$V_K(t^{-1}) \neq V_K(t) \qquad (3.83)$$

then the knot K *is topologically chiral.* (Note, however, that the converse is not always true: equality in the above relation does not necessarily imply topological achirality of the knot K.)

The recursive rules for the construction of the Jones polynomial $V_K(t)$ for an arbitrary knot K are shown in Figure 3.6. By definition, the Jones polynomial $V_U(t)$ of the unknot U is 1,

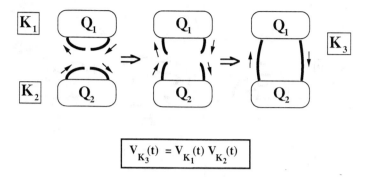

$$V_{K_3}(t) = V_{K_1}(t)\, V_{K_2}(t)$$

Figure 3.7 If two knots K_1 and K_2 are cut anywhere and then are interconnected orientations matching, then the Jones polynomial $V_{K_3}(t)$ of the resulting new knot K_3 is the product of the Jones polynomials $V_{K_1}(t)$ and $V_{K_2}(t)$ of the two original knots K_1 and K_2, respectively.

$$V_U(t) = 1. \tag{3.84}$$

Consider now three knots, denoted by K_-, K_o, and K_+. These three knots are identical under the cover Q, shown in Figure 3.6, and they differ only in their exposed parts, where they have a left-handed crossing, an avoided crossing, and a right-handed crossing, respectively. For any three such knots their Jones polynomials $V_{K_-}(t)$, $V_{K_o}(t)$, and $V_{K_+}(t)$ are interrelated by the equation

$$- t^{-1}\, V_{K_-}(t) - (t^{1/2} - t^{-1/2})\, V_{K_o}(t) + t\, V_{K_+}(t) = 0 \tag{3.85}$$

(Note that here we follow the normalization convention used in an early chemical application [62], where the relation between this normalization and other conventions has been pointed out.)

Another useful rule is illustrated in Figure 3.7. If two knots, denoted by K_1 and K_2, are cut anywhere and then are interconnected orientations matching, then the Jones polynomial $V_{K_3}(t)$ of the resulting new knot K_3 is the product of the Jones polynomials $V_{K_1}(t)$ and $V_{K_2}(t)$ of the two original knots K_1 and K_2:

$$V_{K_3}(t) = V_{K_1}(t)\, V_{K_2}(t). \tag{3.86}$$

In Figure 3.8, an illustration of the application of the recursive relation (3.85) of Jones polynomials is given. For the generation of the Jones polynomial of the right-handed trefoil knot T_+ we proceed in three steps. In Step 1, we start with two unknots U, taking the roles of K_- and K_+, and the third "knot" is the unlinked double ring denoted informally by "U^2", taking the role of K_o. As shown, these three knots can be arranged so that they appear identical under the cover on their left-hand sides and their exposed parts are precisely a left-handed crossing, an avoided crossing, and a right-handed crossing, respectively. The

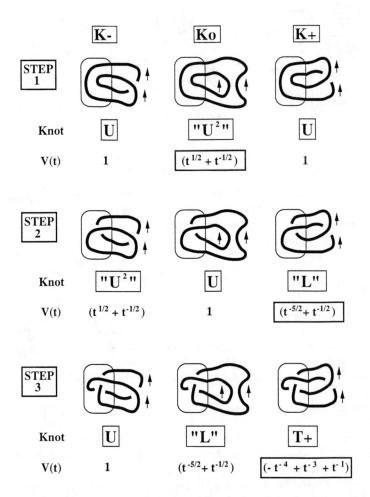

Figure 3.8 An illustration of the application of the recursive relation for the generation of the Jones polynomial of the right-handed trefoil knot T_+. In Step 1, starting with two unknots, the recursive relation gives the Jones polynomial of the unlinked double ring "U^2", shown in the frame. In Step 2, the above polynomial and that of the unknot are used to obtain the Jones polynomial of the linked "engagement rings", "L". In the final Step 3, the polynomials of the unknot U and the engagement rings L result in the Jones polynomial of the right-handed trefoil knot T_+.

conditions of recursive rule (3.85) shown in Figure 3.6 are fulfilled, and from the known polynomial of the unknot U, the recursive rule gives the Jones polynomial $(t^{1/2} + t^{-1/2})$ of the unlinked double ring "U^2", shown in the frame. This result is used in Step 2 by applying the same rule for "U^2" and U, taking the roles of K_- and K_0, respectively, to obtain the Jones polynomial $(t^{-5/2} + t^{-1/2})$ of the linked "engagement rings", informally denoted by "L". In the final Step 3, the polynomials of the unknot U and the engagement rings L result in the Jones polynomial

$$V_{T_+}(t) = (-t^{-4} + t^{-3} + t^{-1}) \qquad (3.87)$$

of the right-handed trefoil knot T_+. The Jones polynomial of the left-handed trefoil knot T_- can be calculated using rule (3.82):

$$V_{T_-}(t) = V_{T_+}\lozenge(t) = V_{T_+}(t^{-1}), \qquad (3.88)$$

that is, for the left-handed trefoil knot T_- the Jones polynomial is

$$V_{T_-}(t) = (-t^{+4} + t^{+3} + t). \qquad (3.89)$$

Clearly,

$$V_{T_-}(t) \neq V_{T_+}(t), \qquad (3.90)$$

the result we expected: we have reconfirmed that the trefoil knots are topologically chiral. In chirality studies, the power of these polynomials lies in their ability to recognize topological chirality even if the knots are very complicated and are arranged in a highly irregular manner with much more actual crossings than their crossing number.

In Figure 3.9, regular projections, the common knot theoretical symbols, and the corresponding Jones polynomials of knots and links of crossing numbers less than six are shown, where the normalization convention of reference [62] is used. Only one topological enantiomer and its Jones polynomial are shown for each pair of topologically chiral knots or links; the Jones polynomial of the mirror image $K^\lozenge$ of each knot K can be easily obtained from the polynomial of the original knot K by replacing the variable t with its reciprocal t^{-1}. The Jones polynomial, as a tool for the detection of topological chirality of a knot can be easily tested for all these examples: topological chirality is implied if the Jones polynomial of a knot is different from the polynomial obtained by substituting t with t^{-1}. Note that topological chirality implies geometrical chirality, but geometrical chirality of a given arrangement does not imply topological chirality, as we have seen in the example of the figure 8 knot.

In fact, stronger and more surprising results are known [71], for example, there are topologically achiral knots which cannot be arranged in a geometrically achiral manner (i.e., they are not "rigidly achiral" according to the knot theoretical terminology). When a geometrically chiral arrangement of such a topologically achiral knot is converted into its mirror image, it is not only possible to do so without ever making it geometrically achiral, but it is impossible to pass through an arrangement that is geometrically achiral. In a later chapter we shall discuss the application of a method of knots and Jones polynomials to a nonvisual analysis of secondary structure and the topological chirality of the folding patterns of long chain molecules, such as DNA and proteins.

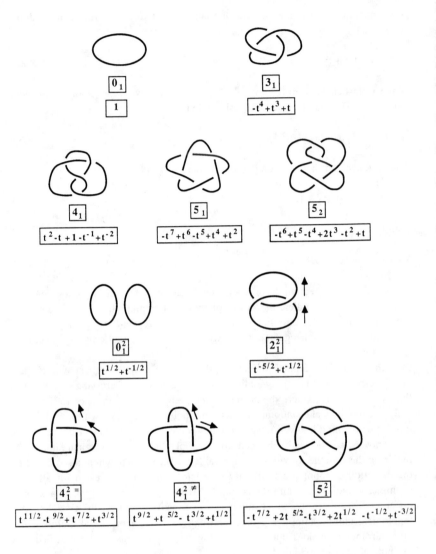

Figure 3.9 Regular projections, the usual knot theoretical symbols, and Jones polynomials of knots and links of crossing numbers less than six. For each pair of topologically chiral knots or links only one topological enantiomer is shown. The Jones polynomial of the mirror image K^0 of each knot K can be obtained from the polynomial of the original knot K by replacing the variable t with its reciprocal t^{-1}. Consequently, the topological chirality of a knot is implied if the Jones polynomial is different from the polynomial obtained by substituting t with t^{-1}. For example, the chirality of the trefoil knot 3_1 and the achirality of the figure eight knot 4_1 are properly reflected if one carries out the above substitution in their Jones polynomials. The method of knots and Jones polynomials can be applied to a nonvisual analysis of secondary structure and the chirality of the folding patterns of long chain molecules, such as DNA, proteins, and chains of synthetic polymers.

3.4   Geometrical Shape and Topological Shape

Geometrical shape is a property of objects which show stable geometrical features and can be characterized in terms of geometry. In practical terms, solid macroscopic objects may be thought of as falling into this category, although no solid object is immune to shape changes due to thermal dilatation, vibrations, and other possible effects. Consequently, geometrical shape is truly an abstraction that can be applied to real objects to various degrees of relevance. However, even if the object is not absolutely stable in terms of geometrical shape, the available geometries (within a time interval, or within a temperature range) often show common, invariant topological features. These features define the topological shape of the object. A good example is provided by knot theory: the geometry of a given knot may show great variations, but its knottedness, exhibiting, for example, topological chirality, is an invariant feature. For objects which preserve their identity under great geometrical variations, such as most molecules, the concept of topological shape is of great importance. Also, as discussed in earlier chapters, the quantum mechanical nature of molecules and the associated consequences of the Heisenberg uncertainty relation imply that localized, geometrical shape characterization can be used only in an approximate sense for molecules, and that a topological description of molecular shape appears more appropriate.

One should not, however, discard the tools offered by geometry. One approach that forms the basis of many of the shape analysis techniques described in later chapters of this book is based on the principle of *geometrical classification and topological characterization*. Geometry is used to define ranges of geometrical arrangements, for example, ranges of allowed distortions, leading to a *geometrical classification* by these ranges, followed by a *topological characterization* and analysis of the invariant topological properties within each range. This principle is used in combination with the shape group methods, described in Chapter 5.

CHAPTER

4

MOLECULAR BODIES, MOLECULAR SURFACES, AND THEIR TOPOLOGICAL REPRESENTATIONS

Molecules are three-dimensional objects and they do occupy some space. When considering the space requirements of molecules, it is natural to associate with them a formal molecular body and a formal molecular surface [84-88]. In a simplistic model, this surface is a formal molecular boundary, a closed surface that separates the 3D space into two parts: the molecular body enclosed by the surface that is supposed to represent the *entire* molecule, and the rest of the 3D space that falls on the outside of the surface, hence on the outside of the molecule. The above, intuitive concepts of molecular body and molecular surface are very useful for the interpretation of molecular size and shape properties within approximate models.

However, real molecules are quantum mechanical objects and they do not have a finite body defined in precise geometrical terms and a finite boundary surface that contains all the electron density of the molecule. The peripheral regions of a molecule can be better represented by a continuous, 3D electronic charge density function that approaches zero value at large distances from the nuclei of the molecule. This density function changes rapidly with distance within a certain range, but the change is continuous. The fuzzy, cloud-like electronic distribution of a molecule is very different from a macroscopic body [251], and no precise, finite distance can be specified that could indicate where the molecule ends. No true molecular surface exists in the classical, macroscopic sense.

Nevertheless, it is possible to construct both classical and approximate quantum chemical molecular models which take advantage of the concepts of a formal molecular body and molecular surface. A formal body and its formal boundary surface can be defined by requiring only that the surface encloses the *essential part* of a molecule. Depending on the chemical problem, there are various choices for what is to be considered the essential part of the molecule. Some of the most commonly chosen approaches will be reviewed in the next section. Contour surfaces of electronic charge densities (MIDCO's), molecular electrostatic potential contours (MEPCO's), contours of molecular orbitals, molecular Van der Waals surfaces generated by fused atomic spheres, solvent accessible surfaces, and various other surfaces surrounding some or all of the nuclei of a molecule can be considered as molecular surfaces and the part of the 3D space enclosed by these surfaces can be regarded as approximate molecular bodies.

4.1 Geometrical and Topological Models for Molecular Bodies and Contour Surfaces

Molecular bodies of quantum mechanical electron distributions or some other molecular functions such as electrostatic potentials can be represented on various levels of approximation. These representations have two main components: the physical property or model used to define a formal molecular body, and the geometrical or topological method used to describe and analyze the model. If a representation of the molecular body is selected, then the boundaries of these approximate molecular bodies can be regarded as formal molecular surfaces. Hence, the molecular shape analysis problem can be formulated as the shape analysis problem of formal molecular surfaces.

Of course, for any given molecule, a single surface cannot provide a detailed enough description of the shape of the actual, fuzzy electron distribution or the entire, 3D molecular electrostatic potential. Often one must consider a whole continuum of a family of molecular surfaces. That is, individual geometrical models are insufficient for the description of the shape of molecules, especially if the conformational flexibility and more general, dynamic molecular properties are considered. Topology can help in two ways: in providing efficient techniques for the shape analysis of individual surfaces, and also as a tool to extract the important, common features from an entire family of such surfaces, which can collectively represent the shape of the molecule. In the above spirit, one can distinguish between the geometrical models of individual molecular surfaces and their enclosed molecular bodies, and the topological models describing the common topological properties of families of contour surfaces and the associated bodies, defined, for example, by a range of electron density contour values.

In this chapter some of the physical properties and approximate models used for molecular shape representation will be reviewed.

4.2   Charge Density and Electrostatic Potential

Electronic charge densities $\rho(\mathbf{r})$ are 3D molecular functions which can be observed experimentally. In fact, X-ray structure determination methods are based on the scattering of X-rays on the electronic cloud of molecules (see, e.g., ref. [89]).

For molecules within periodic crystal lattices, the measured intensity data of the X-rays reflected from the (hkl) plane of the crystal can be converted into the F_{hkl} structure factors, and the observed electronic charge density $\rho(\mathbf{p})$ can be obtained by a technique called Fourier synthesis:

$$\rho(\mathbf{p}) = \sum_h \sum_k \sum_l F_{hkl} \, \exp\left[-2\pi i(hx+ky+lz)\right] , \qquad (4.1)$$

where the coordinates of point $\mathbf{p} = (xa, yb, zc)$ are given with reference to the a, b, and c constants characteristic to the unit cell of the crystal.

Electronic charge densities have fundamental influence on a wide variety of molecular properties. Electron densities are related to the formal sizes of atoms and the formal bond lengths of molecules, for example, in various crystals [278], and there are important relations between experimental electron densities and temperature [279]. Electronic charge densities $\rho(\mathbf{r})$ can be calculated by various quantum chemical methods, both *ab initio* and semiempirical (see, e.g., refs. [90,91]). Density difference calculations are used for direct comparisons of electronic structures (see, e.g., ref. [280]), whereas the effects of electron correlation on charge densities are of special importance in the study of nonbonded interactions [281].

The electronic charge density $\rho(\mathbf{r})$ is a 3D function that can be calculated from the n-electron wavefunction solution $\Psi(\mathbf{r}_1,\mathbf{r}_2,... \mathbf{r}_n)$ of the electronic Schrödinger equation of the molecule:

$$\rho(\mathbf{r}) = n \int |\Psi(\mathbf{r},\mathbf{r}_2,\mathbf{r}_3,... \mathbf{r}_n)|^2 \, d\mathbf{r}_2 d\mathbf{r}_3...d\mathbf{r}_n , \qquad (4.2)$$

where the symbol of integration also implies summation for all spin variables. In most quantum chemical computational studies the electronic charge density is specified in atomic units, a.u., defined as electrons/bohr^3, where the conversion to SI units gives 1 a.u. of electron density = 1.08121×10^{12} C/m^3.

Using an atomic orbital expansion in terms of a set $\{\varphi_i(\mathbf{r})\}$ of normalized atomic basis functions, the electronic density function can be calculated from the relation

$$\rho(\mathbf{r}) = \sum_i \sum_j P_{ij} \, \varphi_i(\mathbf{r}) \, \varphi^*_j(\mathbf{r}) , \qquad (4.3)$$

where P_{ij} is the i,j element of the density matrix $\mathbb{P}$, obtained from an *ab initio* or a semiempirical computation. Note that the accuracy of MIDCO's calculated for different contour density values a is seldom uniform throughout the whole density range, nevertheless, *ab initio* methods with appropriate basis sets are suitable for an adequate representation of the valence shell regions of electron densities in closed shell molecules of moderate size.

If molecular shape is represented by the electronic density, then the shape analysis can be performed on the molecular isodensity contours (MIDCO's) of the calculated density. Some of the elementary properties of MIDCO's have been discussed in Chapter 2.3, with special emphasis on the density domain approach (DDA) to chemical bonding (Chapter 2.4), and on the quantum chemical definition of functional groups of chemistry (Chapter 2.5).

The results of the quantum chemical calculations are dependent on the method used, in particular, the calculated *ab initio* electron densities and the associated MIDCO's are dependent on the basis set used. The electronic charge density is a property more sensitive to basis set variations than energy, although it is energy minimization that leads to the actual approximate wavefunction when using the variational principle within most *ab initio* computational schemes. Some basis sets, in particular the Gaussian basis sets used within the GAUSSIAN family of *ab initio* programs of Pople and coworkers [253], are designed to reproduce valence shell molecular properties. Consequently, when MIDCO's are calculated within such *ab initio* framework, the contours corresponding to the relatively low electron density values in the peripheral valence shell regions of the molecule are more accurate than those corresponding to high electron density values in the core region near the nuclei. In addition, the "cusp condition" of the electronic density is poorly represented by any wavefunction built from Gaussian basis sets, consequently, high-density MIDCO's are often less accurate than those at low densities, even if calculated using basis sets that are not specifically designed for valence shells.

A property that is related to the electronic charge density but provides information often more directly useful in the analysis of interactions between molecules is the molecular electrostatic potential (MEP). Isopotential contours (MEPCO's) are defined analogously to the isodensity contours (MIDCO's), by selecting appropriate constant values   a   for level sets of MEP. The electrostatic potential generated by a molecule has a strong influence on molecular interactions and chemical reactions. The molecular electrostatic potential can be calculated, at least approximately, with little computational effort even for large molecules, and it is frequently used for the representation of molecular shapes and steric interactions between polar regions of biomolecules [92,155,191,282-308]. For any specified conformation of the molecule, the MEP function $V(\mathbf{r})$ generated at point $\mathbf{r}$ can be calculated from the nuclear charges Z_i, the formal nuclear position vectors $\mathbf{R}_i$, and the electronic charge density function $\rho(\mathbf{r})$ as follows :

$$V(\mathbf{r}) = \sum_i Z_i / |\mathbf{r} - \mathbf{R}_i| - \int [\rho(\mathbf{r}')/|\mathbf{r}-\mathbf{r}'|]\, d\mathbf{r}', \qquad (4.4)$$

where the integration in the second term is over the whole space and $\mathbf{r}'$ is the variable of integration. The electronic charge density function $\rho(\mathbf{r})$ and the electrostatic potential $V(\mathbf{r})$ are related by the Poisson equation:

$$\Delta V(\mathbf{r}) = 4\pi\rho(\mathbf{r}) . \qquad (4.5)$$

The atomic unit of electrostatic potential is defined as electrons/bohr, where 1 a.u. of electrostatic potential = 27.2116 V = 3.02770 x 10^{-9} C/m.

One may use multipolar expansions for an approximate calculation of MEP, and it is often convenient to use fractional charges on atoms obtained from *ab initio* or semiempirical quantum chemical MO population analyses [288,300].

A family of molecular surfaces can be defined in terms of the MEP using methods similar to those applied in the case of the electronic density. Since the MEP is a continuous function of the three-dimensional position variable $\mathbf{r}$, it can also be analyzed in terms of level sets $F(a)$ and their contour surfaces $G(a)$ for a selected constant MEP value a, defined analogously to those of charge density :

$$F(a) = \{\ \mathbf{r} :\ V(\mathbf{r}) < a\ \}\ , \qquad\qquad\qquad (4.6)$$

and

$$G(a) = \{\ \mathbf{r} :\ V(\mathbf{r}) = a\ \}\ . \qquad\qquad\qquad (4.7)$$

The above $G(a)$ surface is a molecular electrostatic potential contour surface, MEPCO, for the contour value a. Note that in contrast to the case of electronic charge density contours, in a MEP analysis the function $V(\mathbf{r})$, hence the threshold parameter a, can take both positive and negative values.

Molecules having similar functional groups and those capable of similar reactions, as well as biomolecules and drug molecules of similar biochemical effects often have moieties showing similar MEP contour surfaces, indicating the importance of electrostatic interactions during binding and the initial stages of chemical reactions. The shapes of MEPCO's often suggest mechanistic explanations of how a given ligand interacts with a receptor site of an enzyme. Consequently, the study of the 3D shape of MEPCO's is a useful tool in rational drug design [175,303-308].

The first term in the definition (4.4) of the molecular electrostatic potential is the potential $V_n(\mathbf{r})$ generated by the nuclei, also called the "bare nuclear potential":

$$V_n(\mathbf{r}) = \sum_i Z_i\ /\ |\mathbf{r} - \mathbf{R}_i|\ . \qquad\qquad\qquad (4.8)$$

Parr and Berk [309] have found that the isopotential contours of the nuclear potential $V_n(\mathbf{r})$ of simple molecules show a remarkable similarity to the actual isodensity contours of the electronic ground states of these molecules.

This observation of Parr and Berk provides the basis for a simple approach to molecular shape analysis and molecular similarity analysis, described below. Although the molecular shapes, as defined by the electronic density, differ somewhat from the shapes of the nuclear potentials, their similarity can be exploited: the nuclear potential contour surfaces provide a simple approximation of the shape of molecules. We shall refer to the isopotential surfaces of the nuclear potential contours as NUPCO surfaces. These surfaces have a major advantage: the computation of NUPCO's is a trivially simple task as compared to the calculation of electronic densities. Furthermore, nuclear potential is a useful molecular property in its own right, without any reference to electronic density: a comparison of NUPCO's of various molecules can provide a valid tool for evaluating molecular similarity. The superposition of potentials of *different* sets of nuclei can result in *similar* composite potentials, consequently, the comparison of NUPCO's is better suited for an assessment of molecular similarity than a direct comparison of nuclear arrangements. In particular, the symmetry of a NUPCO is the same as or possibly higher than the point symmetry of the nuclear arrangement. (A higher symmetry is obtained for infinitesimally low potential values.)

A shape analysis of NUPCO's provides a formal "shape signature" of the collective properties of the nuclear arrangement K. Of course, NUPCO's are common for all electronic states of the same nuclear arrangement. Consequently, NUPCO's are not suitable to account for the shape differences between various electronically excited states of molecules, especially in formal "vertical" excitations. However, electronic excitations are often accompanied by changes of the optimum nuclear arrangement K, and NUPCO's can be used for an approximate description of these contributions to the overall shape changes caused by the electronic excitations.

The nuclear potential $V_n(\mathbf{r})$ is a continuous function of the position variable $\mathbf{r}$, as long as $\mathbf{r}$ does not coincide with a nuclear position, $\mathbf{r} \neq \mathbf{R}_i$. The level sets F(a) and their NUPCO boundary surfaces G(a) for any constant potential value a are defined as

$$F(a) = \{ \ \mathbf{r} : \ V_n(\mathbf{r}) < a \ \} , \qquad (4.9)$$

and

$$G(a) = \{ \ \mathbf{r} : \ V_n(\mathbf{r}) = a \ \} , \qquad (4.10)$$

respectively. The value of nuclear potential is positive or zero; there are no NUPCO's G(a) with negative threshold parameter a.

The highest level of shape similarity between NUPCO's and MIDCO's is expected in the high density and high potential core regions near the nuclei, where both contours are essentially spherical. In the valence shell regions of electronic charge density the dominant electron-nuclear interaction is substantially modified by electron-electron interactions, hence one finds greater differences between the shapes of MIDCO's and NUPCO's. Nevertheless, even in the peripheral, valence shell regions, the similarity is significant, and NUPCO's provide a valid approximate model for molecular shapes. A similarity analysis of NUPCO's describes an important aspect of molecular similarity.

A variety of molecular surfaces can be defined in terms of the molecular orbitals $\Psi_i(\mathbf{r})$ of a molecule. A contour surface of an MO $\Psi_i(\mathbf{r})$ is defined in terms of the level sets F(a) and its boundary surface G(a), with respect to a threshold a of the function value $\Psi_i(\mathbf{r})$. Similar to the case of MEPCO's, these threshold values may take both positive and negative values.

The definitions of level sets F(a) and contour surfaces G(a) of an individual molecular orbital $\Psi_i(\mathbf{r})$ are analogous to those of the charge density and electrostatic potential:

$$F(a) = \{ \ \mathbf{r} : \ \Psi_i(\mathbf{r}) < a \ \} , \qquad (4.11)$$

and

$$G(a) = \{ \ \mathbf{r} : \ \Psi_i(\mathbf{r}) = a \ \}. \tag{4.12}$$

The usual *ab initio* Hartree-Fock computations lead to the canonical orbitals which allow a simple assignment of an energy quantity ("orbital energy") to each orbital. In particular, the contour surfaces of the highest (highest-energy) occupied (HOMO) and lowest unoccupied (LUMO) orbitals, and the frontier orbitals [96,97] are often used to interpret properties of molecules and reactions.

The actual sign ("phase") of the molecular orbital at any given point $\mathbf{r}$ of the 3D space has no direct physical significance; in fact, any unitary transformation of the MO's of an LCAO (linear combination of atomic orbitals) wavefunction leads to an equivalent description. Consequently, in order to provide a valid basis for comparisons, additonal constraints and conventions are often used when comparing MO's. The orbitals are often selected according to some extremum condition, for example, by taking the most localized [256-260] or the most delocalized [259,260] orbitals. Localized orbitals are often used for the interpretation of local molecular properties and processes [256-260]. The shapes of contour surfaces of localized orbitals are often correlated with local molecular shape properties. On the other hand, the shapes of the contour surfaces of the most *delocalized* orbitals may provide information on reactivity and on various decomposition reaction channels of molecules [259,260].

4.3   Van der Waals Surfaces and Solvent Accessible Surfaces

Isolated atoms show spherical symmetry, and it is natural to model atoms by spheres of some suitably defined radii. The potential energy of interaction between two atoms rises very sharply at short internuclear distances during atomic collisions, not unlike the potential energy increase in the collisions of hard, macroscopic bodies. In a somewhat crude, approximate sense, atoms behave as hard balls. This analogy can be used for a simple molecular model where atoms are represented by hard spheres. Once a choice of atomic radii is made, the approximate atomic surfaces can be defined as the surfaces of these spheres.

Molecules are built from atoms, and it is natural to relate the formal concept of the "surface of a molecule" to the formal atomic surfaces of the constituent atoms. The radii of atomic spheres chosen for the representation of the space requirements of atoms are usually much too large for modeling molecules by simply placing the atomic hard spheres side by side. This approach would not reflect the internuclear distances correctly, and the surface obtained would have artificial shape features much too different from the shape of the actual electronic charge distribution of the molecule. It is possible, however, to generate various "fused sphere" models for molecules [84-88], by allowing the spheres to interpenetrate one another, setting the distances between the centers of two spheres equal to the formal internuclear distance in the molecule. The spheres can be positioned according to the 3D, stereochemical bond pattern of a particular, fixed nuclear arrangement of the

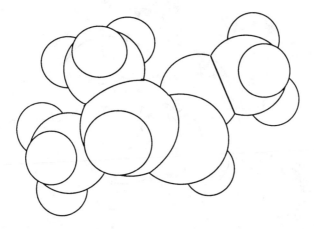

Figure 4.1 An illustration of a fused sphere Van der Waals surface (VDWS) of a molecule.

molecule. The envelope surface of the properly arranged fused spheres may be regarded as a formal molecular surface.

In particular, if the atomic radii are taken as some of the recommended values of the atomic Van der Waals radii, then one obtains a fused sphere Van der Waals surface (VDWS) of the molecule. Several different sets of atomic radii have been proposed [85-87,255], and the fused sphere molecular surface obtained depends on this choice.

The 3D space requirements of most molecules can be represented to a good approximation by such Van der Waals surfaces. Fused sphere VDWS's are used extensively in molecular modeling, especially in the interpretation of biochemical processes and computer aided drug design. These approximate molecular surfaces are conceptually simple, their computation and graphical display on a computer screen take relatively short time, even for large biomolecules.

The usual atomic radii [86] for a VDWS of many molecules provide a good approximation for a MIDCO of an intermediate contour density value a of about 0.002 a.u. One may exploit this fact and design simple VDW representations of molecular surfaces based on molecular charge densities. The methods of fused sphere VDWS's can be extended to generate approximate representations of a family of isodensity contours for a whole range of contour density values [255]. Functions of variable atomic radii, dependent on the desired contour density value have been constructed. These functions allow one to construct simple, scalable, fused spheres surfaces for any desired contour density value a, mimicking MIDCO surfaces G(a) for any a value, using an approximate technique [255] that is computationally much simpler than a direct *ab initio* or even semiempirical calculation of a MIDCO.

In Figure 4.1 an example of a fused spheres Van der Waals surface is shown. This figure illustrates an important difference between a MIDCO and a VDWS: at the seam of interpenetration of the spheres the latter surfaces are not differentiable.

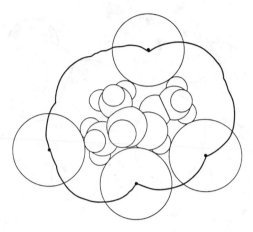

Figure 4.2   A schematic illustration of a solvent accessible surface of the VDWS   of a molecule
shown in Figure 4.1.

Consequently, local curvatures cannot be defined at such points, and the usual
curvature-based partitioning of the shape group method is not applicable (see
Chapter 5.2). This property, however, can be exploited for alternative shape analysis
techniques, where the simplifications offered by spherical surface patches are also
utilized [194,195]. Alternatively, by appropriate "smoothing" these surfaces can be
converted into differentiable surfaces [109], and then the usual shape group methods
are applicable. One such approach is based on minimal surfaces, defined as the
minimum area envelope surface of a set of atomic spheres. This technique is equally
applicable for nonintersecting and fused spheres; the minimal envelope surface of a
VDWS is a natural, smooth extension of the simple fused sphere model of molecular
surfaces. One may picture a minimal envelope surface of a fused spheres VDWS   as
the surface obtained after dipping a model of the VDWS into a soap solution.

    The interactions of a solute molecule with the solvent have important influence
on most molecular properties and reactions. Molecular shape is also affected by such
interactions. One crucial aspect of these interactions is the accessibility of various
regions of the solute molecule by the solvent. A very simple but useful approach is
based on the concept of solvent accessible surface. For the simplest case, the solute
molecule is represented by a fused sphere Van der Waals surface, whereas the
solvent molecule is represented by a sphere. Part of the motivation for this model is
provided by the nearly spherical shape of water MIDCO's at low densities, and by
the fact that solvent molecules are likely to undergo rapid reorientation processes
even within the near vicinity of the solute molecule. An averaging of the various
orientations is expected to result in a nearly spherical time average for many
solvents. Water being the most common solvent, the above model has been used in
many studies of the accessibility of the VDWS   by the solvent molecules [310]. A
simple approach to this problem is based on rolling a spherical solvent molecule on
the VDWS,   and monitoring the positions of the center of the rolling sphere. These

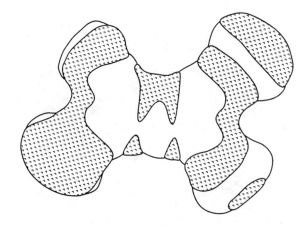

Figure 4.3 An example of an interpenetrating molecular surface pair, such as a MIDCO and a MEPCO pair of the same molecule.

positions define an *envelope surface* about the VDWS, and this envelope surface is taken as a formal solvent accessible surface of the molecule. A solvent accessible surface can be defined for alternative representations of the molecular surface, for example, solvent accessible surfaces can be obtained for various MIDCO's. A special generalization of convexity provides a new representation of such solvent accessible surfaces [262].

In Figure 4.2 a schematic illustration of a solvent accessible surface of the VDWS of Figure 4.1 is shown. A solvent accessible surface also can have points where it is not differentiable. This feature may occur for solvent accessible surfaces of both VDWS' and MIDCO's, even if the latter surfaces themselves are differentiable. As it is shown in Figure 4.2, the rolling solvent sphere can get "bogged down" in local bays along the molecular surface, implying that its smooth motion is interrupted, and it must take a sudden turn. This leads to a discontinuity in the otherwise smooth change of the alignment of the tangent plane of the solvent accessible surface. Such bays can occur along both VDWS and MIDCO surfaces, resulting in nondifferentiable solvent accessible surfaces.

Molecular surfaces representing different physical properties are often markedly different. These differences, as interrelations among various molecular surfaces of the same molecule, can be easily represented by the pattern of interpenetration of two or several such surfaces. The same general technique of interpenetrating surfaces can be applied for two molecular surfaces of the same physical property of two different molecules. In this latter case, the interpenetrating surfaces provide a tool for direct shape comparisons.

For example, the interrelations between the electronic charge density and the electrostatic potential of a molecule can be studied by this technique. An example of such an interpenetrating surface pair is shown in Figure 4.3. A pair of superimposed MIDCO and MEPCO surfaces generates patterns of domains on both surfaces,

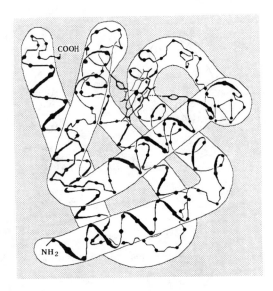

Figure 4.4   A view of the secondary-tertiary structure of myoglobin. The helical segments as well as the heme group are clearly recognizable.

where, for example, a given domain on the MIDCO corresponds to locations where the MEP values are greater then the MEPCO threshold. In general, the shape analysis of these patterns on each of the interpenetrating surfaces can follow the general shape group approach, whereas the envelope surface of the interpenetrating surface pair has points where it is not differentiable.

   One approach for the generation of an appropriate superposition of contour surfaces [311] is based on the distance geometry approaches of Crippen [312], and Crippen and Havel [313]. Most of the software packages used in drug design (see, for example, refs. [314-324]) provide a variety of tools for the related computations. These and related techniques have been applied in medicinal chemistry and in more general molecular engineering problems (see, for example,refs. [325-337]).

   Another family of superimposed surfaces is used in the study of active sites of enzymes. By superimposing approximate molecular surfaces of several molecules showing similar activity with respect to the given enzyme, a part of their envelope surface, called their *union surface,* can be taken as an object that approximately fills out the cavity of the enzyme [167,311,338]. The shape of the union surface (in fact, the shape of its complement) is expected to provide more information on the shape of the enzyme cavity than a surface of a single active molecule.

4.4   Macromolecular Shape and Protein Folding

The shapes of macromolecules, large and complex molecular systems such as polypeptides, proteins, and polysaccharides, can be studied considering different

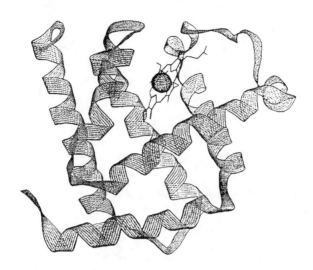

Figure 4.5 A ribbon model of the secondary-tertiary structure of myoglobin, viewed from the same perspective as that of Figure 4.4.

levels of organization. Their molecular arrangements can be viewed at various levels of resolution. Depending on the level of resolution, the relevant shape features can be represented by different types of models, complementing one another. For macromolecules a detailed shape characterization in terms of MIDCO's is often impractical, computationally prohibitively expensive, and in many instances not really necessary. When detailed shape information is required, then formal molecular surface models, such as the fused spheres Van der Waals models are often used to represent a molecule. For macromolecules and long chain molecules the large scale shape features are often more important than the finer details [169,339,340], although detailed shape of some local regions, such as reactive sites, dominant functional groups of macromolecules, and the cavity regions of enzymes are of special importance.

Figures 4.4 and 4.5 show examples of two common types of representations of protein structures. In Figure 4.4 some of the global structural features of the myoglobin molecule are shown from the most commonly used perspective [341]. In this view the helical segments as well as the heme group are clearly recognizable. In Figure 4.5 a ribbon model of the same molecule is shown.

For chain molecules, such as peptides and proteins, the three-dimensional folding pattern is of great importance, and the shape characterization on the corresponding lower levels of resolution requires techniques different from those used for small molecules. For proteins this folding pattern has several, distinguishable levels of complexity. The *primary structural arrangement* of a protein is reflected in the sequence of amino acids. The peptide chain may coil up to form an *α–helix,* or it may form a structure called *β-strand,* whose strands may

Figure 4.6   A simple space curve representation of the central line of secondary structure elements of myoglobin.

combine to form a *β-sheet,* or the chain may remain a *random coil* exhibiting no specific pattern. These patterns represent the *secondary structure* of proteins. The elements of the secondary structure are often represented by simplified symbols, such as the ribbon model of Richardson [169,339,340] and alternative models using, for example, arrows and cylinders [163,172,174,176]. In the ribbon model of Richardson, α-helices, β-strands, and the nonrepetitive loops (or random coils) connecting β-sheets and helices are represented by cylindrical helices of solid ribbons, thick arrows, and thick "ropes", respectively. According to the model of Lesk and Hardman [163,172], the α-helices are modeled by solid cylinders without internal structure, and the β-strands are shown as the arrows of Richardson's model. The relative arrangements of these structural elements of the secondary structure of proteins is often called the *tertiary structure.*

    These models disregard many details of the molecular shape, but they are appropriate to represent the foldings of the macromolecular backbone, the secondary structure of proteins, and the patterns of the tertiary structure.

    A simple 3D representation of the large scale features of a protein is obtained from the ribbon model of Richardson if one considers only the central line of the ribbon. This line is a space curve, usually oriented, from the N-terminal toward the C-terminal of the protein chain. The ribbon model shows the internal turns within a helical segment, consequently, the central line of the ribbon, as a space curve, also shows these details.

    The representation can be further simplified by considering only the central lines of secondary structural entities, α-helices and β-strands as a formal space curve. This curve no longer shows the turns within a helix. Such space curves are used to model the large scale folding pattern of a protein. Note that the random coil segments of the protein (and the entire protein if it is irregular, having no recognizable α-helical segments or β-strands) are represented by the central line of the ribbon or rope as before. As an example, the central line of the myoglobin structure of Figure 4.4 is shown in Figure 4.6. Such space curves can be viewed from many directions and the resulting pattern of the planar projection can show

Figure 4.7   Three views of the space curve representation of the central line of secondary structure elements of the λ - Cro repressor protein.

major variations. In Figure 4.7, three projections of the central space curve of a small irregular protein, the λ - Cro repressor protein are shown. Evidently, the three projections have very different topological patterns. If we are able to characterize all possible planar projections of this space curve, then we have a fairly detailed shape characterization of the folded chain, at least on the level of secondary structural elements. As we shall see in the next chapter, topology, specifically knot theory and graph theory, hold the key to a detailed characterization of the family of patterns of all possible projections [196-198].

An interesting polyhedral representation of helical domains of proteins has been suggested by Murzin and Finkelstein [201]. In their model the helices are modeled by solid cylinders not unlike those in the model of Lesk and Hardman [163,172]. In addition, the random coil segments are modeled by straight line segments. According to their analysis based on a large family of examples, the large scale shape features of most globular proteins are well represented by polyhedra where the edges are the above cylinders and line segments. This technique has been found very useful for protein classification [202], as well as for the analysis of large scale chirality properties of globular proteins [203].

CHAPTER
5

TOPOLOGICAL SHAPE GROUPS, SHAPE CODES, SHAPE GRAPHS, SHAPE MATRICES, AND SHAPE GLOBES

In this chapter we shall combine some of the ideas described in Chapters 3 and 4: the applications of topological concepts and methods for the study of various representations of molecular shapes. Among the shape representations molecular contour surfaces have a prominent role, but we shall also consider alternatives, primarily for the purposes of characterizing the large scale shape features of biological macromolecules.

An important approach to shape analysis is based on generalizations of the concept of convexity. Consider a formal molecular body $B(a)$, taken as the union of a MIDCO $G(a)$ corresponding to a density threshold a, and the level set $F(a)$ enclosed by it. As follows from the definition of convexity discussed in Chapter 3, this body $B(a)$ is a convex set, if for any two points r_1 and r_2 of $B(a)$ all points r of the straight line segment between r_1 and r_2 fall within the body $B(a)$. This is a *global condition* for convexity of a formal molecular body $B(a)$. Such globally convex bodies $B(a)$ represent chemically rather uninteresting electron distributions, since they correspond to either single atoms or to the low electron density MIDCO's of molecules. Due to their simplicity, and to the low density value in the molecular case, their spherical or quasi-spherical shapes are of limited chemical interest. By contrast, nonconvex formal molecular bodies and the associated MIDCO's with more intricate shape features provide more chemically

96

interesting information. For their shape analysis, local convexity and its generalizations are important tools. Some of these generalizations are described in the following sections of this chapter.

Most of the methods discussed in this chapter use the tools of 3D topology. A shape analysis can be based on the two-step process of *Geometrical Classification and Topological Characterization,* and on the principle of *Geometrical Similarity as Topological Equivalence,* where the latter is referred to as the GSTE principle. Geometrical conditions are used to define ranges of geometrical objects (e.g., families of points along a MIDCO where the surface is locally convex) leading to a *geometrical classification* of these points into domains, followed by a *topological characterization* of the various topological properties of the interrelations among these domains.

5.1 Shape Domains of Contour Surfaces

Most molecular contour surfaces defined in terms of various physical properties, such as a MIDCO or a MEPCO, are topologically rather simple objects. Typically, MIDCO's in the molecular density ranges are topologically equivalent to a sphere (when the ordinary, metric topology of the 3D space is used), or in more unusual cases to a doughnut or to a few "fused" doughnuts. The direct topological characterization of such surfaces (using the ordinary, metric topology of the 3D space) does not reveal much detail about their chemically interesting shape features. However, one may use various geometrical or physical conditions, denoted in general by μ, to define some domains D_μ on a contour surface $G(a)$. These domains can be used to subdivide the surface $G(a)$. By cutting out from $G(a)$ all domains of some specified properties [e.g., by eliminating all locally convex domains ("bumps") from the contour surface $G(a)$], a new, topologically more interesting object, a truncated contour surface $G(a,\mu)$, is obtained. This truncated surface $G(a,\mu)$ is no longer topologically equivalent to the original contour surface $G(a)$, nevertheless, $G(a,\mu)$ carries information on the shape of the original surface $G(a)$, where shape is understood within the context of the physical property used to define domains on the surface. A geometrical or physical shape condition is used to turn the original surface $G(a)$ into a topologically different object $G(a,\mu)$, and a topological analysis of the truncated surface $G(a,\mu)$ corresponds to a *shape analysis* of the original molecular surface $G(a)$. The topological invariants of the truncated surface $G(a,\mu)$ contain information on the topological interrelations of various subdivision domains on the original contour surface $G(a)$.

For the characterization of the shapes of molecular contour surfaces, such as MIDCO's and MEPCO's, one may subdivide the surface into domains fulfilling some local shape criteria. One can distinguish two types of criteria, relative, and absolute, leading to a *relative shape domain* or an *absolute shape domain* subdivision of the molecular contour surface.

Relative shape conditions are used when comparing two or several surfaces to one another. For example, a pair of two superimposed contour surfaces of two molecules or of two different physical properties for the same molecule generates an interpenetration pattern on these surfaces (see discussion in Chapter 4), and the maximum connected subsets of this pattern can be taken as the *relative shape domains* on each surface. The relative shape domains in this pattern can serve as criteria for local shape characterization in a relative sense: for one molecule relative to another or for one physical property of a molecule relative to another property. By a topological analysis and characterization of these relative shape domains, a direct comparison is possible between the two molecular surfaces.

An absolute shape characterization is obtained if a molecular contour surface is compared to some standard surface, such as a plane, or a sphere, or an ellipsoid, or any other closed surface selected as standard. For example, if the contour surface is compared to a plane, then the plane can be moved along the contour as a tangent plane, and the local curvature properties of the molecular surface can be compared to the plane. This leads to a subdivision of the molecular contour surface into locally convex, locally concave, and locally saddle-type shape domains. These shape domains are absolute in the above sense, since they are compared to a selected standard, to the plane. A similar technique can be applied when using a different standard. By a topological analysis and characterization of these *absolute shape domains,* an absolute shape characterization of the molecular surface is obtained.

Several topological methods of shape analysis of molecular contour surfaces have been designed to take advantage of such relative and absolute shape domain subdivisions of the contours, according to some physical or geometrical conditions [155-158,199].

For example, the technique of interpenetrating contour surfaces [157] can be applied for a relative shape domain subdivision of a pair of MIDCO and MEPCO surfaces of a molecule. The MIDCO surface can be subdivided into domains using the contour value of the MEPCO as criterion. This procedure is equivalent to generating the interpenetration pattern on the MIDCO surface [157].

One may regard the two interpenetrating surfaces shown schematically in Figure 4.3 as a MIDCO, $G_1 = G_1(a_1)$ and a MEPCO, $G_2 = G_2(a_2)$ of the same molecule. We assume that the contour threshold values are a_1 and a_2, respectively. In general, whether an interpenetration occurs at all depends on the choice of the threshold values a_1 and a_2. If the thresholds are chosen so that an interpenetration does occur, as shown in the figure, then this interpenetration defines one or several closed loops on both surfaces. All points along these loops belong to both contour surfaces, and on each surface the loops define the boundaries of subsets characterized by the function value of the *other* physical property. The value of the given property is either greater than the threshold for all points within the interior of a subset, or it is lower than the threshold for all points within the interior of the subset. Of course, the interpenetration patterns can be generated simultaneously for a series of threshold values, that is, for a series of MEPCO's with respect to a given MIDCO, or *vice versa.* (Considering the interpenetration of a *series* of MIDCO's with a *series* of MEPCO's simultaneously is also a possibility, but if there are several members in both sequences, the patterns obtained may rapidly become intractable.) For example, take k different MEP threshold values, a_2^1, a_2^2, a_2^3,.... a_2^k. The corresponding pattern of interpenetration can be used to generate ranges of MEP mapped onto the MIDCO $G_1(a_1)$. These ranges define a subdivision

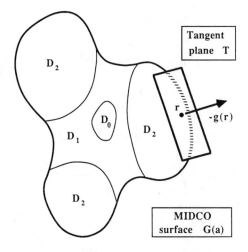

Figure 5.1   The shape domains of local convexity of a MIDCO surface G(a) are shown. A geometrical interpretation of the classification of points **r** of G(a) into locally concave D_0, locally saddle-type D_1, and locally convex D_2 domains is given when comparing local neighborhoods of the surface to a tangent plane T. Each point **r** of G(a) is classified into domains D_0, D_1, and D_2 depending on whether at point **r** a local neighborhood of point **r** on the tangent plane (**r** not included) falls within the interior of the surface G(a), or it cuts into the surface G(a) within any small neighborhood of point **r**, or it falls on the outside of G(a).

of $G_1(a_1)$ into relative shape domains, and the interrelations among these domains can be characterized topologically. We shall see later in this chapter how to generate a simple numerical shape code, which serves as a concise shape characterization of the molecule. This numerical shape code can be evaluated, stored, and compared to that of another molecule by a computer, following the principles of a *nonvisual* approach to molecular similarity analysis [108].

As an example of absolute shape criteria, the local curvature properties of a MIDCO can be used for defining absolute shape domains on it [156], and for a subsequent global shape characterization. In Figure 5.1 a MIDCO G(a) is shown as an illustration of some of the concepts discussed. The simplest method [155] is based on comparisons to a reference of a tangent plane what leads to the identification of locally convex, concave, and saddle-type domains, as mentioned previously, although much finer characterizations are also possible [156,199].

For both relative and absolute shape domains, their topological relations (e.g., their neighbor relations) are invariant within some range of nuclear arrangements (i.e., within some domain of the nuclear configuration space M). The similarities of the shapes of contour surfaces of slightly distorted geometries appear as topological equivalence of their shape domain relations [108,109,155,156,158,199]. The same observation applies for the comparison of molecular surfaces belonging to two different molecules. If the neighbor relations among the corresponding shape domains on the two molecular surfaces are the same (a topological equivalence), then

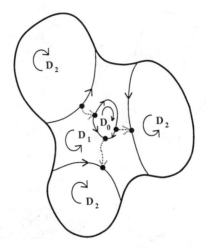

Figure 5.2 A cellular subdivision based on the local convexity shape domains of the MIDCO surface of Figure 5.1 is shown.

the shapes of the two molecular contour surfaces are similar within the above context [108,109,155,156,158,199].

In order to characterize the local curvature properties of a smooth molecular contour surface $G(a)$, one can classify the surface points into curvature types. Imagine that $G(a)$ is the surface of a planet where the ore distribution within the planet causes the force of local gravity to act perpendicular to the surface everywhere. Consequently, the local directions of "up" and "down" are also perpendicular to the surface everywhere. Consider a small neighborhood of each point $\mathbf{r}$ on the surface as a function expressed as "elevation" above the local tangent plane $T(\mathbf{r})$ of $G(a)$ at this point $\mathbf{r}$. This function is positive along the normal vector (of the tangent plane) pointing "up", i.e., away from the interior of $G(a)$, the function is zero at point $\mathbf{r}$, and it is negative "below" the tangent plane. If $G(a)$ is a MIDCO, then this normal vector is the negative gradient vector $-\mathbf{g}(\mathbf{r})$ of the electronic charge density $\rho(\mathbf{r})$, as shown in Figure 5.1. At point $\mathbf{r}$ the molecular surface $G(a)$ and the plane $T(\mathbf{r})$ have a tangential contact with each other, in other words, they *osculate*. Consequently, the gradient of this "function of elevation" (not to be confused with the gradient of the electronic charge density) is zero at $\mathbf{r}$, that is, point $\mathbf{r}$ is a *critical point* of the function of elevation. The matrix of second derivatives of this function, its local Hessian matrix $\mathbb{H}(\mathbf{r})$, expresses the local curvature properties of the contour surface $G(a)$ at each point $\mathbf{r}$. The *eigenvalues* $h_1(\mathbf{r})$ and $h_2(\mathbf{r})$ of the local Hessian matrix $\mathbb{H}(\mathbf{r})$ are the *local canonical curvatures* of surface $G(a)$ at point $\mathbf{r}$ [155,156,199].

If μ denotes the number of negative eigenvalues of the local Hessian matrix $\mathbb{H}(\mathbf{r})$, then point $\mathbf{r}$ is said to belong to a domain D_μ of the contour surface $G(a)$. A local curvature analysis along the surface generates a subdivision into various D_μ curvature domains. For the three possible μ values of 0, 1, and 2, one obtains the locally concave, saddle-type, and convex domains D_0, D_1, and D_2, respectively. These local curvature domains D_μ can be further subdivided, leading to a *cellular subdivision* of $G(a)$. In Figure 5.2, a cellular subdivision based on the local convexity of shape domains of the contour surface of Figure 5.1 is shown. This example will be used in subsequent sections of this chapter, where the shape groups are introduced, and where the conventions for assigning orientations (arrows) to various parts of the surface are also described.

A more general family of methods for absolute shape domain subdivision of molecular surfaces with reference to regular standard objects, such as plane, spheres, and ellipsoids, can be described within the common framework of *generalized convexity* [199]. These techniques are applicable for smooth (differentiable) molecular surfaces.

The local canonical curvatures can be compared to a reference curvature parameter b [156,199]. For each point **r** of the molecular surface $G(a)$ a number $\mu = \mu(\mathbf{r},b)$ is defined as the number of local canonical curvatures [the number of eigenvalues of the local Hessian matrix $\mathbb{H}(\mathbf{r})$] that are less than this reference value b. The special case of b=0 allows one to relate this classification of points to the concept of ordinary convexity. If b=0, then μ is the number of negative eigenvalues, also called the *index* of critical point **r**. As mentioned previously, in this special case the values 0, 1, or 2 for $\mu(\mathbf{r},0)$ indicate that at the point **r** the molecular surface $G(a)$ is *locally concave, saddle-type,* or *convex,* respectively [199].

By generalizing the idea of local convexity for any reference curvature value b [199], the number $\mu(\mathbf{r},b)$ is the tool used for a classification of points **r** of the contour $G(a)$ into various domains. For any fixed b, each point **r** of the contour surface $G(a)$ belongs to one of three disjoint subsets of $G(a)$, denoted by A_0, A_1, or A_2, depending on whether at point **r** none, one, or both, respectively, of the local canonical curvatures h_1 and h_2 are smaller than the reference value b [156]. The union of the three sets A_0, A_1, and A_2 generates the entire contour surface, that is,

$$G(a)=A_0 \cup A_1 \cup A_2 . \tag{5.1}$$

Each of these subsets A_0, A_1 and A_2 of the contour surface $G(a)$ may be disconnected. A maximum connected component of set A_0, A_1 or A_2 is denoted by $D_{0,k}$, $D_{1,k}$ or $D_{2,k}$, respectively, where the first index is the common $\mu(\mathbf{r},b)$ value within the subset, whereas the second index k is simply a serial number of some ordering. Typically, each $D_{\mu,k}$ domain is a two-dimensional subset of the contour surface $G(a)$. In particular,

$$\mathbf{r} \in D_{0,k} \text{ for some } k \quad \text{if and only if} \quad b \leq h_1(\mathbf{r}), h_2(\mathbf{r}), \tag{5.2}$$

$$\mathbf{r} \in D_{1,k} \text{ for some } k \quad \text{if and only if} \quad h_1(\mathbf{r}) < b \leq h_2(\mathbf{r}), \tag{5.3}$$

$$\mathbf{r} \in D_{2,k} \text{ for some } k \quad \text{if and only if} \quad h_1(\mathbf{r}), h_2(\mathbf{r}) < b. \tag{5.4}$$

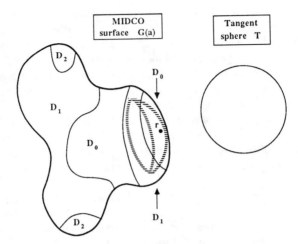

Figure 5.3 The shape domains of relative local convexity of a MIDCO surface G(a) of Figure 5.1, relative to a tangent sphere T of curvature b (radius 1/b) are shown. A geometrical interpretation of the classification of points **r** of G(a) into locally concave D_0, locally saddle-type D_1, and locally convex D_2 domains relative to b is given when comparing local neighborhoods of the surface to the tangent sphere T. The classification depends on whether at point **r** the surface G(a) is curved more in all directions, or more in some and less in some other directions, or less in all directions, than the test sphere T of radius 1/b. In the corresponding three types of domains $D_{0(b)}$, $D_{1(b)}$, and $D_{2(b)}$, or in short D_0, D_1, and D_2, the molecular contour surface G(a) is locally concave, of the saddle-type, and convex, respectively, *relative to curvature* b.

For any fixed curvature parameter b, each point **r** of the molecular surface G(a) belongs to one and only one of the $D_{\mu,k}$ domains of some indices μ and k. Consequently, these $D_{\mu,k}$ domains generate a complete partitioning of contour surface G(a). The mutual arrangements and interrelations of the resulting local shape domains of G(a) give a natural, global shape characterization of the entire surface G(a), where the curvature of the surface G(a) is measured against the given reference curvature b. In the notation used for the $D_{\mu,k}$ domains the reference curvature parameter b is not indicated; note, however, that for a different b value a different set of $D_{\mu,k}$ domains is obtained.

In the above discussion we have assumed that the molecular contour surface G(a) is twice differentiable. This condition is required for gradients and local Hessian matrices of the local elevation function at all points along the surface, and for the local canonical curvatures of G(a) at each point **r** of G(a), needed for their classification into shape domains.

The case of b=0 corresponds to the shape domain subdivision of G(a) in terms of *ordinary local convexity* [155,199]. Geometrically, this case corresponds to comparing the local regions of the molecular contour surface to a test surface of zero curvature, that is, to a tangent plane. Local convexity and the corresponding classification of points **r** of G(a) into various D_μ domains, in the present case into locally concave D_0, locally saddle-type D_1, and locally convex D_2 domains, is illustrated by the example of Figure 5.1 and by the corresponding cellular subdivision shown in Figure 5.2.

The more general case of $b \neq 0$ corresponds to a generalization of the concept of convexity [156,199]. This concept of *relative local convexity* has a useful geometrical interpretation. For a fixed value of parameter b, relative local convexity classifies the points **r** of G(a) into domains $D_{\mu(b)}$, depending whether at point **r** the surface G(a) is curved more in all directions, more in some and less in some other directions, or less in all directions, than a *test sphere* T *of radius* 1/b. The corresponding three types of domains are denoted by $D_{0(b)}$, $D_{1(b)}$, and $D_{2(b)}$, where the molecular contour surface G(a) is locally concave, of the saddle-type, and convex, respectively, *relative to curvature* b.

A classification of points of a MIDCO surface G(a) of Figure 5.1 relative to a tangent sphere T is shown in Figure 5.3, where the reference to parameter b is omitted from the notation. The two shape domain partitionings show important differences; in general, the geometrical pattern of domains varies with a change of reference curvature b. For the entire range of $-\infty < b < \infty$ of the curvature parameter, there are infinitely many geometrical patterns of relative convexity shape domains on any given molecular contour surface G(a). However, for a small change of b, the topological pattern of the shape domains does not change necessarily. In fact, for all but some pathological cases, there are only *finitely many topologically different patterns* of $D_{\mu(b)}$ shape domains of G(a) for the entire range $-\infty < b < \infty$ of curvature parameter b. This is an important observation that brings about a useful simplification: for a detailed topological shape analysis within the above framework of relative local convexity, it is sufficient to consider a finite number of appropriately chosen reference curvature values b. The shape domains of relative local convexity provide the means for studying fine shape features of molecular contour surfaces.

Within the general scheme of relative convexity, the conventional, ordinary local convexity is obtained as a special, degenerate case of relative local convexity, with a tangent sphere of infinite radius as reference, that is, with a tangent plane of reference curvature $b = 0$.

The numerical value of the reference curvature b can be specified in absolute units or in units scaled relative to the size of the object G(a). If absolute units are used, then a relative convexity characterization of G(a) involves size information; if an object G(a) is scaled twofold, then its shape remains the same, but with respect to a fixed, nonzero b value a different relative convexity characterization is obtained. That is, the pattern of relative shape domains $D_{0(b)}$, $D_{1(b)}$, and $D_{2(b)}$ defined with respect to some fixed, nonzero reference curvature value b ($b \neq 0$) is size-dependent. On the other hand, if the reference curvature b is specified with respect to units proportional to the size of G(a), then a simple scaling of the object does not alter the pattern of relative shape domains with respect to the scaled reference curvature b. In this case, the shape characterization is size-invariant, that is, a "pure" shape characterization is obtained.

A natural, size-independent relative convexity characterization is obtained if the relative curvature parameter b is scaled by a size parameter of the object G(a), for example, by the diameter $d(G(a))$. In practice, the radius

$$r(G(a)) = 0.5 \, d(G(a)) \tag{5.5}$$

of the smallest sphere enclosing $G(a)$ is used as an internal reference for scaling the relative curvature parameter:

$$b_G = r(G(a)) \, b \, . \tag{5.6}$$

The relative curvature domains $D_{0(bG)}$, $D_{1(bG)}$, and $D_{2(bG)}$, specified in terms of the scaled reference curvature b_G provide a size-independent shape characterization of the object $G(a)$ for all curvatures.

In an alternative approach, the reference curvature b is scaled by the diameter $d(K)$ of the 3D nuclear configuration K. If $r(K) = 0.5 \, d(K)$ is the radius of the smallest sphere that encloses all the nuclei of the given nuclear configuration K, then the scaled relative curvature parameter b_K is defined as

$$b_K = r(K) \, b = 0.5 \, d(K) \, b \, . \tag{5.7}$$

The resulting curvature domains $D_{0(bK)}$, $D_{1(bK)}$, and $D_{2(bK)}$ are not invariant with respect to the size of the $G(a)$ objects (this size is dependent on the contour parameter a), nevertheless, the scaling is specific for the size of the nuclear arrangement K, hence these shape domains provide a valid shape comparison of MIDCO's or other molecular surfaces of molecules of different sizes. This approach is simpler than the fully size-invariant approach using the reference curvature b_G, where a new scaling factor $r(G(a))$ is required for each new MIDCO $G(a)$.

For the special case of reference curvature $b = 0$ (i.e., for the tangent plane T of ordinary convexity), the pattern of the original curvature domains $D_{0(0)}$, $D_{1(0)}$, and $D_{2(0)}$ is already size-invariant.

As shown in Figure 5.4, the analysis of curvature properties of molecular contour surfaces can be further refined by testing the local curvatures of a contour surface $G(a)$ against an oriented *tangent ellipsoid* T [199]. This technique is designed for shape analysis in external fields or with respect to a direction defined by a nearby molecule. The relative orientation of the reference ellipsoid T and contour surface $G(a)$ is fixed. At each point $\mathbf{r}$ of the surface $G(a)$ the ellipsoid T of axes of fixed orientation can be brought into tangential contact with $G(a)$ by applying a suitable translation of T. Similarly to the case of the tangent sphere, at each point $\mathbf{r}$ the $G(a)$ surface is regarded locally as a function of "elevation" over the tangent ellipsoid T. The second derivatives of this function of elevation define a Hessian matrix $\mathbb{H}_T(\mathbf{r})$, and points $\mathbf{r}$ of $G(a)$ are classified into domains according to the *oriented relative local convexity* properties of their neighborhoods on $G(a)$, *relative to the tangent ellipsoid* T. Points with zero, one and two negative eigenvalues belong to domains $D_{0(T)}$, $D_{1(T)}$, and $D_{2(T)}$, which are concave, of the saddle-type, and convex, respectively, *relative to the oriented tangent ellipsoid* T. In Figure 5.4 two examples of *oriented relative local convexity* shape domain classification of points of the molecular contour surface

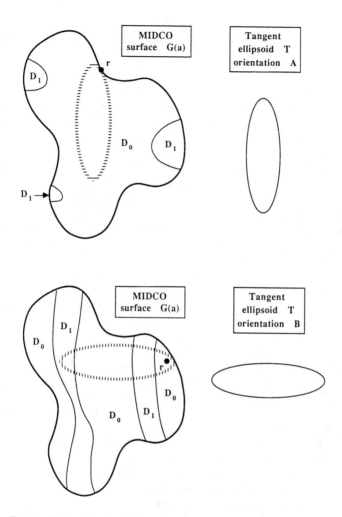

Figure 5.4 Two sets of shape domains of oriented relative local convexity of the MIDCO surface G(a) of Figure 5.1, relative to two orientations of a tangent ellipsoid T are shown.

G(a) of Figure 5.1 are shown, where the subdivisions are given relative to two orientations of a tangent ellipsoid T.

The oriented test ellipsoid T may be chosen to represent an external electromagnetic field, or the main direction of a cavity of an enzyme molecule, or a polarizability ellipsoid of a molecule, or an alignment on the surface of a catalyst, or some other internal or external constraint [199].

In a further generalization of the concept of convexity, the ellipsoid T may be replaced by any other differentiable surface, for example, by a contour surface of another molecule [199]. The resulting shape domains can be used for a direct shape comparison and a direct similarity test for these molecules.

5.2   The Shape Group Method (SGM) for the Analysis of Molecular Shapes

Consider a family of shape domains defined on a molecular contour surface $G(a)$, and the truncated contour surface $G(a,\mu)$ obtained from $G(a)$ by excising a selected subfamily of D_μ shape domains. The *shape groups* of the contour surface $G(a)$, with respect to the given family of shape domains, are the homology groups of the truncated contour surfaces $G(a,\mu)$.

For example, if the shape domains are defined in terms of local convexity, and if we select the locally convex domains, then the shape groups of $G(a)$ are the homology groups of the truncated isodensity contour surface $G(a,2)$, obtained from the molecular contour surface $G(a)$ by eliminating all D_μ domains of index $\mu = 2$. This family of shape groups, obtained by cutting out all locally convex domains of $G(a)$, has been studied in most detail for several molecules [192,262,263,342].

If the shape domains are defined by relative local convexity, then the notation $HP_\mu(a,b)$, $p = 0, 1, 2$, is used for the shape groups of MIDCO surfaces $G(a)$, where besides the dimension p of the homology group, the truncation type μ, the charge density contour parameter a, and the reference curvature parameter b are also specified. For the special case of ordinary local convexity, $b=0$, the second argument in the parentheses can be omitted and one may simply write $HP_\mu(a)$. Usually, we are interested in the Betti numbers of the groups $HP_\mu(a,b)$ and $HP_\mu(a)$; for these numbers the $b_{p\mu}(a,b)$ and $b_{p\mu}(a)$ notations are used, respectively.

It should be emphasized that the above shape group methods combine the advantages of geometry and topology. The truncation of the MIDCO's is defined in terms of a geometrical classification of points of the surfaces, and the truncated surfaces are characterized topologically by the shape groups.

For many chemists, the concept of group theory is intimately connected to molecular symmetry properties. Note, however, that the shape groups are not determined by the point symmetry of the nuclear framework, and these groups give a symmetry-independent characterization of molecular shape.

As an example, we consider the $a=0.01$ MIDCO of the allyl alcohol molecule. Note that a density domain analysis of the bonding and functional groups of this molecule has been discussed in Chapter 2. In Figure 5.5 the $G(0.01)$ MIDCO is shown, where the molecule is oriented the same way as in Figures 2.5 and 2.6. The shape of this MIDCO can be characterized by its shape groups corresponding to the shape domains of relative local convexity. In Figure 5.6 the shape domains corresponding to three different choices of curvature parameter b are shown. The shape domains corresponding to $b=0$ are

- two simply connected domains of type D_2 [one on the far side of $G(a)$],
- one D_2 domain with two holes in it (near the nuclei of the OH group),
- one D_2 domain with three holes in it (near the vinyl moiety, top left),
- three simply connected domains of type D_1 (the smallest near the H of OH),
- and one D_1 domain with four holes in it.

Following the computation of the homology groups as described in Chapter 3, the $\mu=2$ type truncation leads to the family of shape groups $HP_2(0.01)$, $p = 2, 1, 0$, of

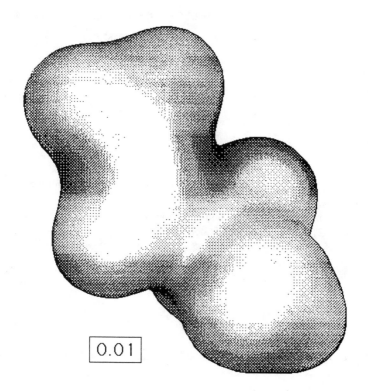

Figure 5.5 The G(0.01) MIDCO of the equilibrium nuclear configuration of the allyl alcohol molecule is shown, where the molecule is oriented the same way as in Figures 2.5 and 2.6.

Betti numbers

$b_{2,2}(0.01) = 0,$

$b_{1,2}(0.01) = 3,$

and

$b_{0,2}(0.01) = 4.$

A shape domain partitioning in terms of relative local convexity of parameter b=0.005 leads to a simpler pattern. We obtain
- one simply connected D_2 domain (near the OH group),
- two D_2 domains each with two holes [one such D_2 on the far side of G(a)],
- two simply connected D_1 domains [one on the far side of G(a)],
- and one D_1 domain with three holes in it.

The shape groups $HP_\mu(0.01,0.005)$, $p=2,1,0$, of the $\mu=2$ type truncation have the following Betti numbers:

$b_{2,2}(0.01,0.005) = 0$,

$b_{1,2}(0.01,0.005) = 2$,

and

$b_{0,2}(0.01,0.005) = 3$.

The third shape domain partitioning shown has been calculated for the relative local convexity parameter $b= - 0.008$. We obtain
- eight simply connected D_2 domains [one on the far side of $G(a)$],
- and one D_1 domain with eight holes in it.

The shape groups $HP_\mu(0.01,-0.008)$, $p=2,1,0$, of the $\mu=2$ type truncation have the following Betti numbers:

$b_{2,2}(0.01,-0.008) = 0$,

$b_{1,2}(0.01,-0.008) = 7$,

and

$b_{0,2}(0.01,-0.008) = 1$.

In the above example of Figure 5.6 only three of the topologically different relative shape domain patterns of the MIDCO $G(0.01)$ of the equilibrium nuclear configuration of the allyl alcohol molecule are shown. For all three of these patterns, their topological properties do not change if the curvature parameter b is changed by a small amount: the corresponding shape groups and their Betti numbers are invariant within ranges of the curvature parameter b. However, the topology of the pattern can change for larger variations of b. Nevertheless, there are only a finite number of different sets of shape groups which occur for this MIDCO for the entire range $- \infty < b < \infty$ of the curvature parameter b. Consequently, a finite set of Betti numbers of the finitely many shape groups of the MIDCO provide a detailed shape characterization.

A similar consideration applies if one changes the density threshold $a = 0.01$ of the MIDCO. For most small variations of the density threshold parameter a the shape groups of the corresponding MIDCO's stay invariant. For the entire range of $0 < a < a_{max}$ of the density threshold value a, there are only a finite number of possible shape groups for the given molecule of a fixed nuclear configuration. Furthermore, if one considers limited deformations of the nuclear arrangement K, for example, by taking the chemical identity preserving deformations within a catchment region of a potential energy surface, then, again, only finitely many shape group types H^1_2 of the actual shape groups $H^1_2(a,b,K)$ may occur. In the above

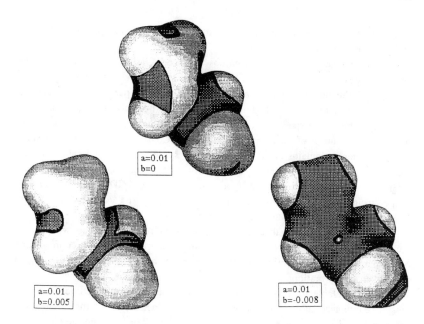

Figure 5.6 Three of the topologically different shape domain patterns of the G(0.01) MIDCO of the equilibrium nuclear configuration of the allyl alcohol molecule are shown, corresponding to reference curvature parameter values b=0, b=0.005, and b=-0.008, respectively.

notation the nuclear configuration K is also specified. The shape of the entire molecular charge distribution, with all its allowed deformations, can be characterized in detail by a finite number of shape groups.

We can formulate the above ideas more precisely by considering the dynamic shape properties of molecules within a nuclear configuration space M.

Our interest is to determine what shapes are present within each catchment region, which shapes are similar and how the shapes change during conformational changes and chemical reactions. This leads to the study of the shape group distributions in the nuclear configuration space M. As it has been discussed in Chapter 2.2, for a general N nucleus system (N≥3) a configuration can be specified by 3N-6 internal coordinates. However, for a dynamic shape analysis of MIDCO surfaces in terms of their shape domains of relative local convexity and the associated shape groups, some additional parameters are also of importance: the electron density threshold value a along the MIDCO surface G(a), and the reference curvature b of the tangent sphere T. For a detailed description of the dynamic shape properties of the molecule, a range of the 3N-6 internal coordinates, and a range of the two parameters, a and b, are needed. One may consider parameters a and b as formal, additional coordinates augmenting the 3N-6 internal coordinates of the nuclear arrangement. The full set of these coordinates defines a formal space, the *dynamic shape space* D of 3N-6+2 = 3N-4 dimensions [158]. Each point of D corresponds to a formal, fixed nuclear arrangement, and

to a pair of a specified density contour value a   for the corresponding MIDCO
and a reference curvature value b.

As discussed above, the shape groups $H^1_2(a,b,K)$ are locally invariant to most
small changes in the coordinates of the dynamic shape space D, that is, to small
changes of the nuclear configuration K, the contour parameter a, and the
reference curvature b. The shape groups do change but only at exceptional points
of the dynamic shape space D (in a formal mathematical sense, these exceptional
points form a subset of measure zero within D). Consequently, a family of shape
group invariance domains of the dynamic shape space D can be assigned to each
shape group type, for example, to the most commonly used 1D homology groups
$H^1_2(a,b)$, obtained for the $\mu = 2$ type truncation of the MIDCO G(a). For each
shape group type H^1_2, these invariance domains of the actual shape groups
$H^1_2(a,b,K)$ generate a partitioning of the dynamic shape space D.

In molecular shape analysis it is important to address the question, what shapes
are available for a given molecule? By regarding a nonrigid molecular conformation
of a separate chemical identity (of some energetic stability) as the catchment region
of a given conformational minimum on the potential surface, one may ask what
shapes are available for this conformation? When addressing these questions, one
must take into account the dynamic properties of the species, and all the variations of
the formal molecular geometry which are allowed while preserving the identity of
the species.

Due to the construction of the dynamic shape space D, the nuclear
configuration space M is one of its subspaces. Consequently, it is meaningful to
refer to projections of parts of D onto M, just as it is meaningful to project a part
of the 3D space onto a 2D plane. Such projections can be used for shape
comparisons. One may ask the following question: which shape group invariance
domains of the dynamic shape space D have projections on a given catchment
region C(0,i) of the nuclear configuration space M? This is to ask, what shapes
(as described by shape groups) are available for a given chemical species? The
projection of a part of D onto a catchment region C(0,i) of M can be visualized
to occur along the coordinates a   and b [158]. One may expect that the following
trend is true: the more shape domains have projections on a given catchment region
C(0,i), the larger number of different reactions involve the corresponding chemical
species [108].

Consider two different subsets of the same space D, or subsets of two
dynamic shape spaces D and D' of two different stoichiometric families of
molecules. One may compare those domains of the two subsets that belong to the
same shape group H^1_2. Since within these domains the nuclear configuration is not
fully specified, that is, there exists some configurational freedom within these
domains, the above approach provides a description of the *dynamic similarity* of
molecular shapes. We shall return to the problems of dynamic shape similarity in
Chapter 6.

For a given nuclear arrangement K the shape group distribution as a function
of the two parameters a and b defines an (a,b)-map for each shape group type.
These maps show the domains of the a,b parameter plane where the given shape
group type, such as the 1D homology group $H^1_2(a,b)$ obtained for the $\mu = 2$ type

truncation of the MIDCO G(a) is invariant. In a formal sense, the shape analysis can be iterated: one can characterize the shapes of these planar invariance domains of the given shape group type. The (a,b)-map of the 1D shape group $H^1_2(a,b)$ of the $\mu = 2$ truncation gives a detailed shape characterization of the entire family of MIDCO's G(a) for the full range of $0 < a < a_{max}$ of the density threshold parameter a, and for the full range of $-\infty < b < \infty$ of the relative curvature parameter b.

Each (a,b)-map can be regarded as a subset of the dynamic shape space D. Such a subset contains all points of D where the internal coordinates corresponding to the nuclear arrangement are fixed.

In practice, it is often sufficient to consider a single shape group type for a specified type of truncation. Then, one may also consider each separate piece of the MIDCO's obtained in the truncation process as a separate entity with its own shape group. An example for such a modified (a,b)-map has been calculated for the allyl alcohol molecule.

In Figure 5.7, a representation of the (a,b)-map of the shape groups of the equilibrium configuration charge density of the allyl alcohol molecule is shown, as calculated with a 6-31G* basis set. Since one has to consider a very wide range of parameter values, it is advantageous to use a logarithmic scale for both parameters a and b. In the case of relative curvature parameter b that can take both positive and negative values, the log|b| value is considered. According to the convention used, the lower half of the logarithmic map corresponds to negative b values. At the level of resolution used for this logarithmic (a,b)-map, all low absolute values of curvature parameter b below 10^{-5} are formally compressed to the horizontal line at log|b|= - 0.5. Consequently, this line corresponds to the case of ordinary convexity, i.e, to the case of tangent plane of zero curvature, b=0. (This tangent plane can be regarded as a reference sphere T of infinite radius.) To consider all shape groups of all dimensions and all possible truncation types for each (a,b) pair is inconvenient, consequently, here only the one dimensional shape groups are specified for the $\mu = 2$ type truncation.

For further convenience, instead of considering the entire object obtained after truncation as a single entity, here each separate piece of the truncated MIDCO surface is regarded as a new object, and the shape group is specified for each separate piece. Accordingly, each surface piece of each truncation (for each choice of density a and curvature b) is characterized by Betti numbers. This approach is suitable for a better identification of local shape features, especially important in studies of large molecules and in shape complementarity analysis, discussed in Chapter 6. However, the number of separate pieces of the surface is not a constant: if the values of the a and b parameters change, the number of separate surface pieces can also change. As a consequence of this approach, the number of Betti numbers assigned to one point of the (a,b)-map can be different from the number of Betti numbers assigned to some other points of the map. When numerical shape codes are generated from such (a,b)-maps, some care must be taken to account for the changes of the number of Betti numbers. According to one of the simplest options used for a concise shape characterization, families of Betti numbers associated with a given location of an (a,b)-map are encoded by a single numerical code. Using a numerical key, the actual Betti numbers of the various surface pieces can be recovered.

In the example shown in Figure 5.7, a legend is provided, showing the actual correspondence between the various families of Betti numbers and the numerical codes used as entries in the (a,b)-map. Each one of the largest families of Betti numbers of the example contain four members, corresponding to four separate surface pieces obtained after truncation. These families are (0,0,0,0), (0,0,0,2), (0,0,0,4), and (0,0,0,3), encoded as 6, 7, 13, and 15, respectively. The smallest families each contain just one member; these single Betti number families are (0), (3), (4), (7), (8), (6), (9), (5), and (10), encoded as 2, 11, 14, 16, 18, 19, 20, 21, and 22, respectively. These single member families correspond to truncations resulting in a single surface piece. There are three families with two Betti numbers and four families with three Betti numbers.

Note that in special cases, two negative integers may replace the families of Betti numbers. If the truncation eliminates the entire MIDCO, a formal no-group situation, this is indicated by the numerical symbol -2. If at the given level of resolution an (a,b) point falls on the borderline of two shape group domains of the map, then the symbol -1 is used to replace the family of Betti numbers. In our example, these two special cases are encoded as 1 and 5, respectively.

The choice for the single number codes used as entries in the (a,b)-map may follow an arbitrary convention. The convention used in the example of Figure 5.7 is based on the frequency of occurrence of the families of Betti numbers and the special "no group" and "borderline" cases.

In Chapter 6, alternative and more descriptive coding techniques will be discussed, where a formal "shape identity number" or a "shape identity vector" can be assigned to objects. The encoding and decoding of these shape identity descriptors are somewhat cumbersome for all but the smallest molecules. The advantage of the method illustrated in Figure 5.7 is the fact that a simple numerical key is used. Each numerical symbol gives a short-hand notation for the corresponding set of the Betti numbers of the one-dimensional shape groups of the separate pieces of the truncated surface, or for the case of -2 of a "no-group" situation, or for the case of -1 when the shape group assignment is ambiguous at the given level of resolution. The encoding-decoding steps are simple, however, the code itself, without the key, is not sufficient for a reconstruction of the shape information. It is possible to construct only slightly more complex numerical codes where the first entry specifies the size of the legend table, followed by the legend table and a grid of entries of the actual (a,b)-map. A shape code based on this principle can be decoded without additional information.

For the study of most intermolecular interactions, valence shell properties, and for practical applications in drug design, the lower density MIDCO's are more important than those at high threshold values. Consequently, the [0.001,0.1] density interval for the a values usually provides sufficient information for shape comparisons. Furthermore, a finite grid on the (a,b) map appears satisfactory for shape characterization. In some recent applications [263], a grid of 41 x 21 = 861 points have been used, taking 41 values from the above density interval and 21 values from the [-1,+1] interval for reference curvature value b.

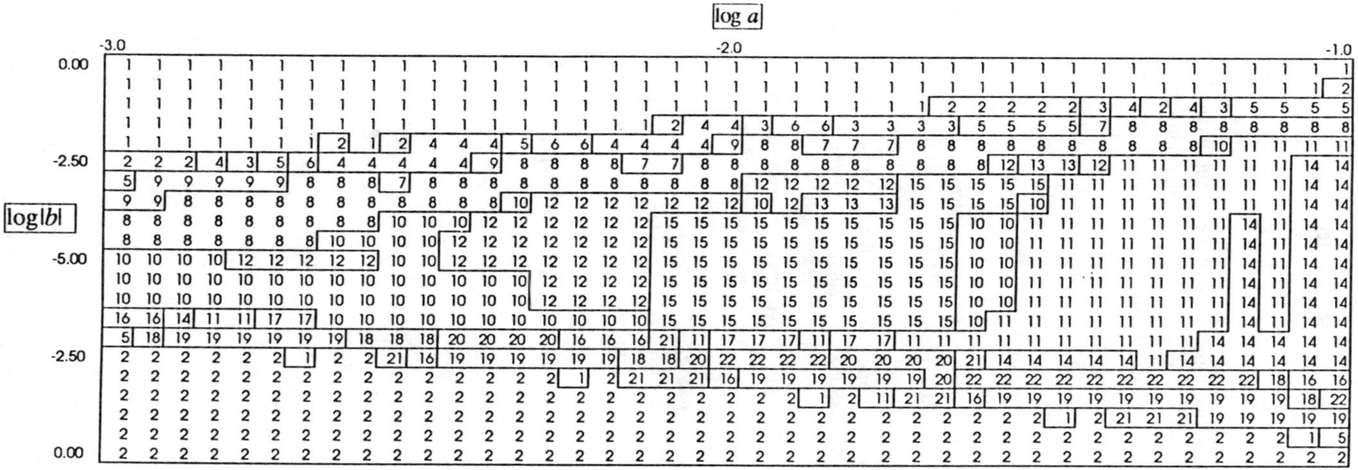

Figure 5.7 A representation of the (a,b)-map of the shape groups of the charge density of the allyl alcohol molecule at equilibrium nuclear configuration is shown, as calculated with a 6-31G* basis set. Logarithmic scale is used for both parameters a and b. Curvature b can take both positive and negative values, hence the log|b| value is used. The lower half of the map corresponds to negative b values. All low absolute values of b (below 10^{-5}) are compressed to the horizontal line at log|b| = - 0.5. The one-dimensional Betti number is specified for each separate piece of the truncated MIDCO, where the $\mu=2$ type truncation is used. The numerical symbol -2 indicates that the truncation eliminates the entire object, a no-group situation, whereas -1 indicates that the given (a,b) pair falls on the borderline of two shape group domains of the map. Otherwise, the Betti numbers are listed for each separate piece obtained in the truncation. The lists of Betti numbers correspond to single number entries in the (a,b)-map, as shown in the legend table.

Legend

1 = -2	12 = 0 0 4
2 = 0	13 = 0 0 0 4
3 = 0 0 0	14 = 4
4 = 0 0	15 = 0 0 0 3
5 = -1	16 = 7
6 = 0 0 0 0	17 = 0 3
7 = 0 0 0 2	18 = 8
8 = 0 0 2	19 = 6
9 = 0 2	20 = 9
10 = 0 0 3	21 = 5
11 = 3	22 = 10

5.3   Shape Codes, Shape Graphs, and Shape Matrices

After the points of a molecular surface are classified into convex, concave and saddle-type domains using some relative local curvature properties, the relations of these domains can be characterized by the topological shape group methods. These shape groups generate a numerical shape code: one can collect the Betti numbers of the corresponding shape groups into a vector or matrix, and this vector or matrix provides a numerical shape characterization. Alternatively, the mutual arrangements and interrelations of the shape domains can be characterized directly by their neighbor relations, leading to another matrix, the *shape matrix* [109,110,158,193], or to an equivalent graph representation of the shape, using the *shape graph* [109,110,158] of the given MIDCO.

Consider a MIDCO $G(a)$ and a choice for the curvature parameter b, and assume that the shape domains $D_{\mu,k}$ of relative convexity of $G(a)$ have been determined. By using an appropriate neighbor relation to describe the mutual arrangements of the $D_{\mu,k}$ domains along the MIDCO surface $G(a)$, the corresponding *shape matrix* $s(a,b)$ and the associated *shape graph* $g_\mu(a,b)$ can be defined [109,110,158,193].

Two $D_{\mu,i}$ domains are considered N-neighbors if they have a common boundary line. For a more precise description, the N-neighbor relation between two $D_{\mu,i}$ shape domains is defined in terms of their closures $clos(D_{\mu,i})$. In accord with the definitions given in Chapter 3, the closure $clos(D_{\mu,i})$ of a domain $D_{\mu,i}$ contains all the points of $D_{\mu,i}$ as well as all of its boundary points. The formal definition of the N-neighbor relation is given below:

$$N(D_{\mu,i}, D_{\mu',i'}) = \begin{cases} 1 & \text{if } (clos(D_{\mu,i}) \cap D_{\mu',i'}) \cup (D_{\mu,i} \cap clos(D_{\mu',i'})) \neq \varnothing \\ \\ 0 & \text{otherwise.} \end{cases} \qquad (5.8)$$

This neighbor relation is similar to the "symmetric strong neighbor relation" between some potential surface catchment regions of reaction topology, used in the analysis of reaction mechanisms [106,343-345].

For the MIDCO surface $G(a)$ of a given nuclear arrangement and for a selected shape domain partitioning relative to curvature parameter b, the *shape matrix* $s(a,b)$ is defined [109,110,158,193] as follows:

$$s(a,b)_{i,i'} = N(D_{\mu,i}, D_{\mu',i'}), \quad i \neq i', \qquad (5.9)$$

and

$$s(a,b)_{i,i} = \mu_i, \qquad (5.10)$$

where μ_i is the common $\mu(\mathbf{r},b)$ index for all points $\mathbf{r}$ within the i-th shape domain $D_{\mu,i}$.

As long as the assignment of the i indices to the various $D_{\mu,i}$ domains is arbitrary, any two shape matrices $s(a,b)$ and $s'(a,b)$ related to one another by simultaneous row and column permutations describe equivalent shapes. One may, however, choose the assignment of index i to follow the ordering of the $D_{\mu,i}$ shape domains according to the size of their surface areas on the MIDCO $G(a)$. According to one convention [109], one may list them according to decreasing size. If such an ordering is chosen, then the shape matrix $s(a,b)$ encodes both shape and size information. In this case, the comparison of molecular shapes (and to some extent, sizes) can be reduced to a comparison of their shape matrices. The problem of molecular shape comparison, a task conventionally requiring visual inspection, is converted to a matrix comparison, a task that computers can perform without requiring direct human involvement.

As our first example, the rather simple shape matrix $s(a,b) = s(a,0)$ of the shape domains of ordinary local convexity of the MIDCO surface of Figure 5.1, is given below:

$$
s(a,b) \;=\;
\begin{array}{ccccc}
1 & 1 & 1 & 1 & 1 \\
1 & 2 & 0 & 0 & 0 \\
1 & 0 & 2 & 0 & 0 \\
1 & 0 & 0 & 2 & 0 \\
1 & 0 & 0 & 0 & 0 .
\end{array}
\qquad (5.11)
$$

In this matrix the index ordering follows the order of decreasing size of the surface area of the various $D_{\mu,i}$ domains.

If a different reference curvature value b is chosen, then the shape matrix may be different, although the matrix is invariant within small enough intervals of the b values. As examples, the three different shape matrices $s(0.01,0)$, $s(0.01,0.005)$, and $s(0.01,-0.008)$ of the three shape domain partitionings of the allyl alcohol MIDCO $G(0.01)$ shown in Figure 5.6 are given below. The index ordering of the various $D_{\mu,i}$ domains follows the order of decreasing size of their surface area:

$$
s(0.01,0) \;=\;
\begin{array}{cccccccc}
2 & 1 & 0 & 1 & 0 & 0 & 1 & 0 \\
1 & 1 & 1 & 0 & 1 & 1 & 0 & 0 \\
0 & 1 & 2 & 0 & 0 & 0 & 0 & 1 \\
1 & 0 & 0 & 1 & 0 & 0 & 0 & 0 \\
0 & 1 & 0 & 0 & 2 & 0 & 0 & 0 \\
0 & 1 & 0 & 0 & 0 & 2 & 0 & 0 \\
1 & 0 & 0 & 0 & 0 & 0 & 1 & 0 \\
0 & 0 & 1 & 0 & 0 & 0 & 0 & 1 ,
\end{array}
\qquad (5.12)
$$

$$s(0.01,0.005) = \begin{matrix} 2 & 0 & 1 & 0 & 1 & 0 \\ 0 & 2 & 1 & 0 & 0 & 0 \\ 1 & 1 & 1 & 1 & 0 & 0 \\ 0 & 0 & 1 & 2 & 0 & 1 \\ 1 & 0 & 0 & 0 & 1 & 0 \\ 0 & 0 & 0 & 1 & 0 & 1, \end{matrix} \tag{5.13}$$

and

$$s(0.01,-0.008) = \begin{matrix} 1 & 1 & 1 & 1 & 1 & 1 & 1 & 1 & 1 \\ 1 & 2 & 0 & 0 & 0 & 0 & 0 & 0 & 0 \\ 1 & 0 & 2 & 0 & 0 & 0 & 0 & 0 & 0 \\ 1 & 0 & 0 & 2 & 0 & 0 & 0 & 0 & 0 \\ 1 & 0 & 0 & 0 & 2 & 0 & 0 & 0 & 0 \\ 1 & 0 & 0 & 0 & 0 & 2 & 0 & 0 & 0 \\ 1 & 0 & 0 & 0 & 0 & 0 & 2 & 0 & 0 \\ 1 & 0 & 0 & 0 & 0 & 0 & 0 & 2 & 0 \\ 1 & 0 & 0 & 0 & 0 & 0 & 0 & 0 & 2. \end{matrix} \tag{5.14}$$

If the entire range of curvature parameter b is considered, then a list of the finite number of distinct shape matrices and those curvature values b_j where a change of the shape matrix occurs, gives a detailed, numerical shape characterization of the MIDCO surface $G(a)$. In the most general case of variations in the two parameters a and b, as well as in the nuclear configuration K, one can study the dynamic shape space invariance domains, the (a,b)-maps, and various projections of the invariance domains of shape matrices, following the principles [158] applied for the shape group invariance domains of the dynamic shape space D.

In general, a *graph* is fully specified if the entities considered as the vertices of the graph, and the pairs of the vertices which are connected by edges are defined [159]. Any square matrix m with offdiagonal elements equal to either zero or one can be thought of as a representation of a graph. In this graph the vertices correspond to the diagonal elements $m_{i,i}$ of the matrix so that each vertex is labeled by the corresponding diagonal element, and there is an edge between the vertices corresponding to matrix elements $m_{i,i}$ and $m_{j,j}$ if and only if the offdiagonal element $m_{i,j} = 1$.

For the *shape graph* $g(a,b)$ of MIDCO $G(a)$, the *vertex set* is the family of $D_{\mu,i}$ domains [158]:

$$V(g_\mu(a,b)) = \{ D_{\mu,i} \}, \tag{5.15}$$

and the *edge set* of the shape graph $g(a,b)$ is the family of pairs of $D_{\mu,i}$ domains

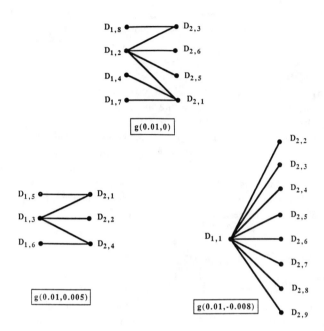

Figure 5.8 Three shape graphs, $g(0.01,0)$, $g(0.01,0.005)$, and $g(0.01,-0.008)$, of the three shape domain partitionings of the allyl alcohol MIDCO $G(0.01)$ of Figure 5.6 are shown. These shape graphs correspond to the three shape matrices $s(0.01,0)$, $s(0.01,0.005)$, and $s(0.01,-0.008)$, given by Equations (5.12), (5.13), and (5.14), respectively.

with nonzero N-neighbor relation:

$$E(g(a,b)) = \{(D_{\mu,i}, D_{\mu',i'}): N(D_{\mu,i}, D_{\mu',i'}) = 1\} \ . \tag{5.16}$$

Both the shape matrix $s(a,b)$ and the shape graph $g(a,b)$ give a detailed shape characterization of the MIDCO surface $G(a)$, with respect to the selected reference curvature b.

In Figure 5.8, three shape graphs, $g(0.01,0)$, $g(0.01,0.005)$, and $g(0.01,-0.008)$ are shown. These shape graphs correspond to the three shape domain partitionings of the allyl alcohol MIDCO $G(0.01)$ of Figure 5.6, that is, to the three shape matrices $s(0.01,0)$, $s(0.01,0.005)$, and $s(0.01,-0.008)$, given by Equations (5.12), (5.13), and (5.14), respectively.

In most cases, nonzero N-neighbor relations are found for the $(D_{0,i}, D_{1,i'})$ and $(D_{1,i'}, D_{2,i''})$ types of pairs of D_μ domains. Note that each $D_{\mu,i}$ domain is a maximum connected component of the set A_μ of all points of $G(a)$ with index μ, consequently, a nonzero N-neighbor relation is impossible between two D_μ shape domains of the same index μ.

A nonzero N-neighbor relation between a D_0 and a D_2 domain is possible in exceptional cases. A point $\mathbf{r}$ of the MIDCO surface $G(a)$ can simultaneously belong to a $D_{0,i'}$ domain and to the closure $clos(D_{2,i})$ of a $D_{2,i}$ domain,

$$\mathbf{r} \in clos(D_{2,i}) \cap D_{0,i'}, \qquad\qquad\qquad (5.17)$$

if and only if at this point $\mathbf{r}$ both of the local canonical curvatures $h_1(\mathbf{r})$ and $h_2(\mathbf{r})$ are equal to the reference curvature b. This implies a $(D_{2,i}, D_{0,i'})$ edge within the shape graph $g(a,b)$ of the given MIDCO.

We have seen that a simple list of Betti numbers of the shape groups can serve as a *numerical shape code* for a partitioned molecular surface. Some of the alternative topological shape descriptors of molecular surfaces, such as the shape matrices $s(a,b)$ and shape graphs $g(a,b)$, can also serve as 3D topological shape codes [43,109,110,158,199]. In Chapter 6, several examples of shape codes are described and used as numerical shape similarity measures.

5.4   Shape Globe Invariance Maps   (SGIM)

It is natural to imagine molecular shape properties as they would appear to an observer moving about a sphere enclosing the molecule. If the observer is able to characterize all possible views, this characterization can provide a detailed shape description.

A simple approach that may be considered is to project a molecular image (the chosen molecular model or a shape descriptor P) onto a spherical surface, assuming a light source in the center of a sphere and regarding the sphere as a screen. This approach leads to a two-dimensional representation of the molecule on a spherical surface. However, each point of the spherical image contains only local shape information and the projected image is not suitable for distinguishing features that are present in multiply folded patterns along a molecular surface where the light beam passes through several of these folds before reaching the spherical screen. In such cases, various side views are more revealing, where features not seen along the radial lines of the chosen sphere can also be monitored. A technique that overcomes these problems is employed in the method described below.

The general method of Shape Globe Invariance Maps (SGIM, [196]), and its special case first employed for protein backbones [112,197,198,346], are based on a spherical map drawn around the molecular model where each point of the map is characterized by those shape characteristics of the molecular model which are "visible" from the given point of the sphere. This is analogous to what actually happens when one studies the shape of an object like a potato: holding the potato in one's hand and turning it around, viewing it from many different directions, and trying to find all the significant shape features of the potato. Each viewing direction corresponds to a picture, and the totality of all these pictures for all viewing directions characterizes the shape of the potato. In addition, if the potato is enclosed in a sphere, then each viewing direction also corresponds to a point of the sphere, hence one can assign the picture observed from a given direction to the corresponding point of the sphere. This fundamental principle is applicable to a wide

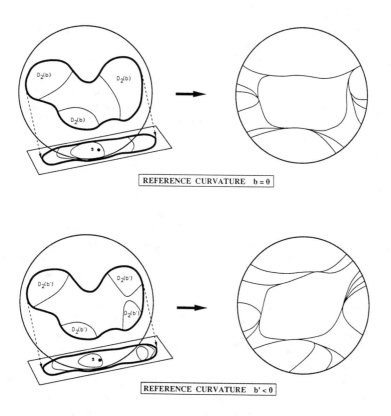

Figure 5.9 The construction of Shape Globe Invariance Maps (SGIM's), of MIDCO relative convexity shape domain patterns for two reference curvature values, b = 0 and b < 0.

variety of primary molecular models and the associated shape representations P, taken as the object viewed from the surface of the sphere. The possible shape representations P include the relative convexity domain partitioning of a MIDCO surface with respect to some reference curvature parameter b, the pattern of interpenetration of two or several molecular isoproperty surfaces (e.g., MEP ranges on a MIDCO surface), space curves representing the backbone of biopolymers ([112,196-198], and references therein), ribbon models of the pattern of protein structural motifs [169,339], or polyhedral models of the folding of helical domains of proteins [201-203].

A practical implementation of the above approach is the following: a global shape property of the molecule is assigned to each point of the sphere S, followed by the determination of those domains of S where this shape property is invariant. A pair of examples is shown in Figure 5.9, where the shape globe invariance domains of a MIDCO surface for two relative convexity shape domain partitionings (P) with respect to two reference curvatures, b = 0, and b' < 0, are given. As before, the molecular shape representation P is enclosed within a sphere S. For example, one can take the smallest possible sphere S that contains the entire shape representation P, provided that the center of mass of the molecule is placed so that it coincides with the center of the sphere. Instead of projecting the molecular descriptor P onto the spherical surface, project P onto a tangent plane $T(s)$ at each point s of the sphere. The projection $P'(s)$ of the shape representation P in each tangent plane $T(s)$ (e.g., each projected pattern of various D_μ shape domains of a MIDCO, or the crossing pattern of a protein's backbone as visible from the given point s) can be characterized topologically, leading to a family of topological descriptors

$$F_j(s) = \{I(i), i=1,...k\}. \tag{5.18}$$

Usually, along the spherical surface one can find several different families of such topological descriptors, and the index j is used to distinguish them. For example, if P is chosen as the local relative convexity domains of a MIDCO surface $G(a)$ of the molecule, then the topological pattern of the 2D image of the corresponding curvature domain partitioning of $G(a)$ can be assigned to each point s of S, as it is projected to the tangent plane $T(s)$ of S at point s. A given family $F_j(s)$ of topological descriptors, assigned to a point s of the sphere, remains invariant within some domain C_j of the sphere. These projected *shape invariance domains* C_j on the sphere S are analogous to countries on a global map, hence the pattern they generate on the sphere is called a *shape globe invariance map* or simply a *shape globe map.*

One should note that within each shape globe map an entire family of topological descriptors $F_j(s) = \{I(i), i=1,...k\}$ is assigned to each point s of the sphere S, providing information on a *global shape property* of the enclosed molecule.

In the examples of Figure 5.9, one tangent plane and the corresponding projection is shown for each reference curvature. The topological pattern $F_j(s)$ of this projection is assigned to the point of tangent s on the shape globe S. Different points s with the same topological pattern $F_j(s)$ are collected into shape globe invariance domains C_j of the sphere S.

The topological descriptors $F_j(s)$ can be chosen in a variety of ways. One approach is a direct application of the shape group method to the shape globe S itself. Truncations of certain domain types of the projected planar image define various shape groups of the 2D image, and these shape groups may be chosen as the topological descriptors $F_j(s)$ ultimately assigned to points s. In this case, an invariance domain C_j of the shape globe map is a maximum connected component of the collection of all points s of S where the shape groups $F_j(s)$ of the images of local relative convexity domains projected on the tangent planes $T(s)$ are the same. Alternatively, neighbor relations of the projected local relative convexity domains define shape matrices of the 2D images, and these matrices (or the corresponding shape graphs) may be regarded as the topological descriptors $F_j(s)$ assigned to points s. In this case, the same shape matrix is assigned to every point of an invariance domain C_j of the shape globe S. Note that the SGIM may

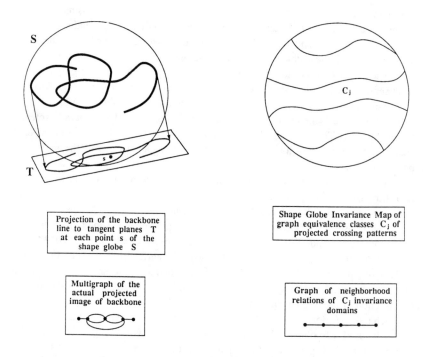

Figure 5.10 The two main steps for the generation of an SGIM based on the crossing pattern graph of the folding of a protein backbone line are shown. The first main step is the projection of a protein backbone line to all possible tangent planes of a shape globe S and the generation of the graphs of the projected crossing patterns associated with each tangent point **s**. These graphs, which may happen to be multigraphs (having more than one edge between some vertices) or pseudographs (having edges starting at and returning to the same vertex) are taken as the shape descriptors $F_j(s)$. In the second main step, the invariance domains C_j of these graphs are generated on the shape globe S. The resulting SGIM on the spherical surface S is a two-dimensional characterization of the three-dimensional shape of the molecular backbone.

appear more complicated than the original curvature domain pattern on the MIDCO. However, a SGIM also takes into account the projected silhouette of the MIDCO, that represents information on all curvature ranges simultaneously. Consequently, the SGIM contains more shape information than the given topological D_μ pattern.

Special cases are discussed in some detail in the literature [112,197,198], where the shape representation P is chosen as a space curve representing a protein backbone and the topological descriptors $F_j(s)$ on the local tangent plane projections are either graphs or knots defined by the crossing pattern on the planar projection at each tangent plane T(**s**) of the sphere S.

An example of this approach is shown in Figure 5.10, where the shape representation P is the space curve representing a protein backbone, and the shape descriptors $F_j(s)$ are chosen as the graphs generated by the overcrossing pattern in the projected image of this backbone. Each crossing, as well as the endpoints of the projected image correspond to a vertex of the graph, whereas the edges of the graph are the projected line segments interconnecting these vertices.

These shape descriptors may turn out to be objects that are not graphs in the strict sense: they may happen to be multigraphs, having more than one edge between some vertices, or pseudographs, having edges starting at and returning to the same vertex [159]. For sake of simplicity in the terminology, we shall generally refer to them as graphs. The resulting SGIM on the shape globe S is a two-dimensional characterization of the global, three-dimensional shape features of the molecular backbone.

Whereas the above representations of three-dimensional shape are two-dimensional, they appear on the surface of a sphere, that is a slight inconvenience when compared to the easier analysis of planar shapes. For this reason, it is useful to construct a planar representation of the shape globes, using a rather simple method. The construction of a planar representation P(SGIM) of a shape globe invariance map SGIM of a MIDCO surface is illustrated in Figure 5.11. Note that this P(SGIM) itself can be regarded as a new shape representation of the molecule, replacing the original shape representation P used for the construction of the original shape map SGIM.

The shape globe S with a shape globe invariance map is placed within a hemisphere H of radius twice that of shape globe S, with a single common point of contact with H and with a plane parallel with the perimeter of H, as shown in Figure 5.11. For simplicity, only one invariance domain C_j of the SGIM is indicated in the figure. The "North Pole" **n** of the sphere S is the point diametrically opposite to the point of contact with the plane. From point **n**, a line is issued to each point **s** ≠ **n** of the shape globe S, piercing the hemisphere H at a unique point **h**. A second line issued from point **h** perpendicular to the plane defines a unique point **p** of the plane. Repeating this procedure for each point **s** ≠ **n** of the shape globe S, an assignment of the points of the shape globe S to points of the plane is obtained. Consider now a modified shape globe S from where the point **n** at the "North Pole" is removed. This object, denoted as S**n**, is a *punctured sphere*. The above pair of consecutive projections is an assignment

$$f : S\backslash n \rightarrow D \hspace{5cm} (5.19)$$

between the punctured shape globe S**n** and an open disk D of the plane. This assignment f is a *bijection*, that is, the assignment of points is one to one in both directions.

One can extend this assignment for the entire shape globe S: the perimeter of the hemisphere H, and as a consequence, the perimeter of the open disk D are assigned to the north pole **n** of the shape globe S. Note that this extended assignment is no longer a bijection. This completes the generation of a planar representation P(SGIM) of the entire SGIM of the shape globe S. An algorithmic shape analysis of such planar representations is not fundamentally different from that

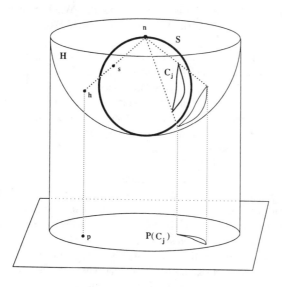

Figure 5.11 The constructions of a planar representation P(SGIM) of a shape globe invariance map SGIM of a MIDCO is shown. The shape globe S with an SGIM is placed within a hemisphere H of radius twice that of shape globe S, over a plane parallel with the perimeter of H, as shown. For simplicity, only one invariance domain C_j of the shape globe invariance map is indicated. From the "North Pole" **n** of sphere S a line is issued to each point $s \neq n$ of S, piercing H at a unique point **h**. A second line issued from point **h** perpendicular to the plane defines a unique point **p** of the plane. This procedure defines an assignment of the points of the shape globe to points of the plane. The shape globe S with its point **n** at the "North Pole" removed, denoted as S\n, is a *punctured sphere*. The above assignment f : S\n → D between the punctured shape globe S\n and an open disk D of the plane is a *bijection* (i.e., the assignment of points is one to one in both directions). The perimeter of the hemisphere H, and as a consequence, the perimeter of the open disk D are assigned to the north pole **n** of the shape globe S. This completes the generation of a planar representation of the entire SGIM of the shape globe S.

of a spherical representation, however, for visual inspection the planar maps are more suitable.

The planar representations P(SGIM) of the shape globe maps SGIM can also be characterized topologically, for example, by their shape groups as defined by a specified truncation pattern (e.g., by eliminating projected invariance domains of certain types) or by the neighbor relations of the projected invariance domains on the planar map P(SGIM). The latter method leads to a treatment analogous to the shape matrix and shape graph methods. The information on the size of invariance domains on the planar map P(SGIM) can be encoded by the ordering of the domains, just as it is done for ordinary shape matrices, discussed above. Such shape matrices provide alternative numerical shape codes, based on the SGIM and P(SGIM) approaches.

5.5   Shape Analysis of Fused Sphere Van der Waals Surfaces and Other Locally Nondifferentiable Molecular Surfaces

Fused sphere surfaces, such as fused sphere Van der Waals surfaces (VDWS') are simple approximations to molecular contour surfaces. By specifying the locations of the centers and the radii of formal atomic spheres in a molecule, the fused sphere surface is fully defined as the envelope surface of the fused spheres and can be easily generated by a computer. Although fused sphere VDW surfaces are not capable of representing the fine details of molecular shape, such surfaces are very useful for an approximate shape representation.

The nondifferentiability of these surfaces at the seams of interpenetrating spheres as well as the local nondifferentiability of solvent accessible surfaces or union surfaces, are a technical disadvantage. Local nondifferentiability limits the application of the shape group methods in their original form that requires second derivatives for curvature analysis. For example, at every point r of a VDWS where two or more atomic spheres interpenetrate one another, the surface is not smooth and is not differentiable. For such nondifferentiable molecular surfaces, alternative shape descriptors and shape codes have been introduced.

One such shape descriptor is based on the minimum number of interior points enclosed by the surface from where the entire interior wall can be illuminated, or, in an equivalent formulation, from where the entire interior wall can be seen. A *seeing graph* [347] for any closed surface, whether differentiable or not, has the following properties:

1. any point of the interior of the closed surface is "seen" by at least one vertex of the graph,
2. each pair of vertices that "see" each other are connected by an edge of the graph, and
3. the graph has the smallest number of vertices with the above properties.

The sequence of seeing graphs for families of MIDCO's of the ethanol molecule has been used for shape characterization [347], and the method is equally applicable to fused sphere VDW surfaces, and to solvent accessible surfaces.

In an alternative shape characterization of VDW surfaces [195,348-350] the various spherical faces of the surface are distinguished depending on the number of circular arcs on their perimeter. A single, separate sphere is regarded as a "0-type face", both spherical faces of the VDW surface of HF molecule are "1-type faces", and on many VDW surfaces "2-type", "3-type", and "4-type" faces are common. By selecting a truncation criterion, for example, by removing all "1-type" faces, and taking the homology groups of the truncated surface for shape characterization, the shape group methods are applicable [195,348-350].

In another alternative based on the neighbor relations and mutual arrangements of the various parts of the atomic spheres which are exposed on the VDW envelope surface, graphs and matrices analogous to the shape graphs and shape matrices are obtained [43,110]. Furthermore, by "smoothing" the VDWS near the interpenetration lines of fused spheres using appropriate "belts" which join smoothly the spherical surfaces, the curvature domain partitioning and the topological methods of similarity analysis of differentiable surfaces are directly applicable. Smooth surfaces developed from fused sphere VDWS models [109] retain some of the conceptual simplicity of a VDWS; one of the suggested smoothing techniques is based on minimal envelope surfaces of fused spheres, as discussed in Section 4.3.

One technique which is applicable for surfaces that are not everywhere differentiable is also suitable for the shape characterization of dot representations of molecular surfaces such as the Connolly surfaces [87], which are not only nondifferentiable, but are not even continuous. The method of T-*hulls* [351] is based on a generalization of the concept of *convex hull*. The convex hull of a set A is the smallest convex set that contains A. Consider a three-dimensional body T. The T-hull of a point set A is the intersection of all rotated and translated versions of T which contain A. The T-hull method is suitable for shape comparisons with a common reference shape, chosen as that of the body T. Alternatively, when the shapes of two molecules, T and A are compared, one molecular body can be chosen as T and the T-hull of the other molecular body A provides a direct shape comparison [351].

5.6 Dynamic Shape Analysis: Topological Principles

As discussed in Chapter 1, the concept of molecular shape has important dynamic aspects, and by contrast to many classical, macroscopic objects, molecular shape is not a static property. The inherent vibrational and other internal motions of molecules, and on a more fundamental level, the quantum mechanical uncertainty in nuclear positions imply that molecular shapes cannot be described in detail without taking into account dynamic aspects.

On the simplest level, one can consider a semiclassical model of limited motions of various parts of the molecule relative to one another. Within such approximation, the dynamic shape variations due to internal motions, for example, those due to vibrations, can be modeled by an infinite family of geometrical arrangements. Within this approach, we consider a family of shapes occurring for these arrangements and study the common, invariant topological features.

As it has been pointed out in Section 5.2, it is natural to formulate dynamic shape analysis aproaches in terms of the dynamic shape space D described earlier [158]. The reader may recall that the dynamic shape space D is a composition of the nuclear configuration space M, and the space of the parameters involved in the shape representation, for example, the two-dimensional parameter space defined by the possible values of the density threshold a, and the reference curvature parameter b of a given MIDCO surface.

We shall distinguish two types of methods for dynamic shape analysis. The methods of the first type are used to determine which nuclear arrangements are associated with a given topological shape. The methods of the second type determine the available topological shapes compatible with some external conditions, for example, with an energy bound.

Within the simplest formulation of a dynamic shape analysis method of the first type, the invariance of topological descriptors within domains of the dynamic shape space D is exploited. The subsets of the dynamic shape space with a common shape group, that is, the shape group invariance domains of D, can serve as tools for a dynamic shape analysis. Within these subsets a limited change of nuclear configurations, hence a limited change in the geometrical shape of the MIDCO surface is permitted, yet these changes are small enough so that within the given topological context the topological shape remains invariant (i.e., the shape group is preserved). The same principle is applicable for other tools of topological shape description, such as the shape matrices, shape graphs, and SGIM's of a given molecule. The dynamic shape space invariance domains of SGIM's serve as tools for analyzing dynamic shape properties.

In one example of a dynamic shape analysis method of the second type, a family of nuclear arrangements is selected, using an upper limit for energy as criterion. By identifying those invariance domains of topological descriptors in the dynamic shape space D which incorporate these nuclear configurations, an energy-dependent family of allowed shapes is obtained, as defined by the given topological descriptors. One may replace the energy criterion with formal temperature, using properties of Boltzmann distributions, for example, by assigning the average energy to each formal temperature value. At a higher temperature, that is, if more energy is available, the molecular vibrations may cover a wider range of formal molecular geometries, hence a greater variety of dynamic shapes occur. At an even higher temperature, where the energy is sufficient for overcoming the activation barriers to conformational rearrangements, a further (usually more significant) increase in the extent of shape variations is found.

Evidently, the dynamic shape of molecules is an energy-dependent property: the changes in the conformational freedom of molecules at various temperatures imply a temperature dependence of molecular shapes. For a "cold" molecule with energy not much exceeding the zero-point energy associated with the various vibrational modes, only a limited choice of possible nuclear arrangements (nuclear configurations) can occur with significant probability. Consequently, only limited shape variations are allowed and the dynamic shape of the system is strongly constrained. By contrast, if the molecule has energy much above the zero-point energy, then it can access a much larger family of possible nuclear configurations with significant probability, and the dynamic shape of the molecule is less restricted. The energy dependence of the accessible shapes [248] and the accessible symmetries [247] of various molecules suggests a family of rules influencing the mechanism and outcome of chemical reactions.

Methods of the second type for the dynamic shape analysis of molecules can also be formulated within the shape globe invariance map framework. The dynamic shape space invariance domains of SGIM's serve as tools for analyzing dynamic shape properties. Clearly, if conformational rearrangements of the nuclei occur, then a given point s on the sphere S may correspond to different families of topological descriptors before and after the change of nuclear arrangement. After the conformational change is completed, the point s may be relabeled as a member of a different invariance region on the sphere S. One approach involves assigning specific labels to those points changing their allegiance between shape globe invariance domains in the course of the conformational change and considering these "no man's land" areas on the shape globe as new, separate domains. The topological characterization of the shape globe may follow the steps described in the previous section. The shape matrices of the resulting dynamic shape globe maps are shape codes including some information on the dynamic shape properties of the molecule.

5.7 Chain Molecule Shape Graphs and Shape Polynomials

In this section we shall describe two approaches to the shape characterization of the large-scale features of chain molecules: one based on a graph-theoretical method, the other on a family of knot theoretical polynomials [112,197,198].

The chosen primary representation of the large-scale features of the chain molecule is a smooth space curve, describing, for example, the central line of rods representing the helical domains of proteins and their interconnecting random coil segments. We may think of this space curve,

$$\mathbf{r}(t), \ 0 \leq t \leq 1, \tag{5.20}$$

as being parametrized by the scalar parameter t, where t=0 and t=1 correspond to the N-terminal and the C-terminal ends of the protein backbone, respectively. As a simple characterization, one can consider three orthogonal projections of this curve to planes, for example, to three planes, each perpendicular to one of the three axes of inertia of the molecule. The results of these projections are three plane curves,

$$\mathbf{q}_i(t), \ 0 \leq t \leq 1, \ i=1,2,3. \tag{5.21}$$

These planar curves may show various crossings, and the pattern of these crossings provides a simple shape characterization of the planar curves $\mathbf{q}_i(t)$, i=1,2,3, and as a consequence, of the space curve $\mathbf{r}(t)$.

The crossing patterns can be characterized by graphs. For each plane curve $\mathbf{q}_i(t)$, i=1,2,3, a *graph* g_i is defined as follows. The vertices of the graph g_i are the two endpoints and the crossover points of the projected backbone curve $\mathbf{q}_i(t)$. (Note that in degenerate cases entire line segments may be projected on one another; in such a case, each maximum connected component of the planar set covered more than once by the projected image is regarded as a vertex, in addition to the endpoints.) The vertices are numbered according to their occurrence when moving along the curve, from t=0 to t=1. The edges of the graph g_i are the projected line segments interconnecting the vertices. In most cases the resulting construction g_i will not be a graph in the strict sense [159], since multigraphs as well as pseudographs are likely to occur. We recall that in a multigraph more than one edge can connect two vertices, and in a pseudograph an edge may start and end at the same vertex. Nevertheless, for simplicity in the terminology, we refer to g_i as a graph.

Since a natural orientation is assigned to the curves, from t=0 to t=1, the crossings can be characterized as right handed or left handed, following the standard

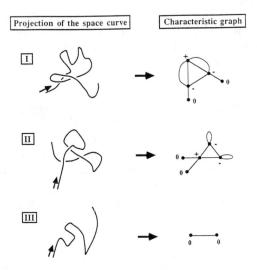

Figure 5.12 Examples of the characteristic graphs of the crossing patterns of three orthogonal views of the chain molecule backbone of the λ-Cro Repressor protein. The chain is oriented from the N-terminal to the C-terminal and the vertices of the graphs are labeled according to the handedness of the crossing: - for left handed crossings, + for right handed crossings, 0 for the two terminals of the chain (no crossing).

convention shown in Figure 3.4. These crossings can be used as vertex labels for the graphs. If the j-th vertex of the graph g_i is denoted by v_{ij}, then its crossing label C_{ij} is defined as follows:

$$C_{ij} = \begin{cases} 1, & \text{if the vertex corresponds to a right-handed crossover,} \\ -1, & \text{if it corresponds to a left-handed crossover, and} \\ 0, & \text{if it corresponds to an endpoint of the curve.} \end{cases} \qquad (5.22)$$

Note that vertices of degenerate cases can be distinguished by an appropriate label. Also note that in earlier works [112,197,198] a somewhat different convention was used, where the endpoints were omitted. If the graph g_i has n vertices, then these labels can be stored in an n-dimensional crossing vector $\mathbf{C}_i$:

$$\mathbf{C}_i = \mathbf{C}_i(g_i) = (C_{i1}, C_{i2}, \ldots, C_{in}). \qquad (5.23)$$

The crossing vector is well defined as long as the projections of all the crossings are regular.

As an illustration, in Figure 5.12 we consider the example of a small irregular chain molecule, the λ-Cro Repressor protein, discussed in refs. [112,197]. The space curve corresponds to a model with structureless helices where only the central line of helices is considered. Such simple molecular space curves show the large-scale arrangement of the folding pattern but not the details of the elements of

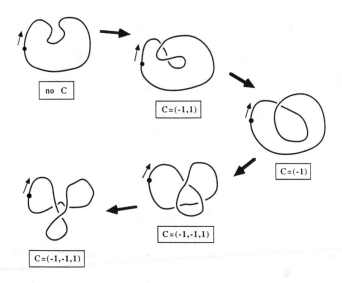

Figure 5.13 A topological characterization of the progress in a folding process of a molecular backbone (represented by a space curve), using vectors **C** of crossing indices. For comparisons with knot theoretical characterizations, a closed curve is chosen. In order to obtain a closed loop for open-ended chain molecules, the two ends can be formally closed by a straight line segment. If in a special projection of the molecular backbone one endpoint is projected on the top of the other, then their interconnecting line is perpendicular to the viewing plane, and the line segment is represented by a dot on the projections, as shown. Otherwise, two dots are used to mark the endpoints of the straight line segment. (An alternative convention is described later for knot generation from open chain molecular models.) The sequence of crossing vectors belonging to various stages of the molecular rearrangement provides information on shape changes during the folding process.

the secondary structure (helices, β-strands, and loops). On the left-hand side of the figure, three orthogonal projections are shown, and the corresponding characteristic graphs g_1, g_2, and g_3 are given on the right hand side. On these graphs the labels +1 and -1 are given by symbols +, and -, respectively. These three graphs characterize the essential folding pattern as viewed from the three chosen directions.

In a conformational change of a chain molecule, for example, in a continuous folding of a protein backbone, the various projections can change, leading to a change in both the graph and the set of crossing labels. However, these graphs and labels remain invariant along segments of the path of the conformational change, and typically there are only a finite number of labeled graphs in each folding process. The sequence of graphs and the sequence of crossing vectors (the vertex labels of these graphs) can be used to characterize the large-scale features of the folding process.

As an example, in Figure 5.13 a topological characterization of five stages in a folding process of a molecular backbone (represented by a space curve) is shown, using vectors **C** of crossing indices for topological characterization. In order to facilitate comparisons with alternative, knot theoretical characterizations, a closed curve is chosen. In one of the simplest representations, a special projection of the molecular backbone is chosen where one endpoint is projected on the top of the other. For open-ended chain molecules, the two ends can be formally closed by a straight line segment. If the special projections are chosen, then the interconnecting line is perpendicular to the viewing plane, and the line segment is represented by a dot on the projections, as shown in the figure. The sequence of crossing vectors belonging to various stages of the molecular rearrangement provides information on shape changes during the folding process. An alternative convention for generating closed loops to model chain molecules will be described in the following. This alternative convention is used for the construction of knots and the associated polynomials.

In Chapter 3 some aspects of knot theory have been described. One may recall that a mathematical knot K (not to be confused with a nuclear configuration K) is a closed space curve in 3D, where the "degree and type of knottedness" can be characterized by various projections of the curve onto planes and by the corresponding crossing pattern. Here we shall use a polynomial characterization of chain molecule folding patterns, based on the Jones polynomial $V_K(t)$ of a knot K representing the molecular backbone. These polynomials provide a nonvisual shape characterization of curves in the 3D space, hence they are useful tools in the computer-based algorithmic comparison of space curves representing the backbone structure of chain molecules.

Knot theoretical techniques are easily applicable to polymer chains that do form actual knots or links, such as some DNA fragments or various catenanes [59-72,204-213]. By appropriate modifications, the knot theoretical polynomials are also applicable to the analysis of chirality properties of general molecules that may not form knots by themselves, but the space around them can be represented by a knot. This approach has led to the concept of *chirogenicity*, and to a nonvisual, algorithmic, computer-based analysis of molecular chirality [62].

An alternative technique is used for general chain molecules [197,198]. Most biologically important chain molecules do not form knots or links, yet knot theoretical methods are applicable for their shape characterization. One can take a projection of the molecular backbone and characterize the projection (where for simplicity, we shall assume that all crossings of the projection are nondegenerate). For example, the space curve of the median line of a ribbon model of a protein is not in general a knot, since the two endpoints of the median line are usually not joined. Nevertheless, for the given projection we may convert the space curve of the median into a knot K_a by the following steps:

1. Attach to each endpoint of the molecular space curve a straight line segment, perpendicular to the viewing plane and pointing away from the viewer. If these line segments are long enough then they must reach a plane that is parallel with the viewing plane and lies beyond the most distant point of the original space curve.

2. Join the far ends of these line segments by another straight line segment within the second plane.

This procedure converts the molecular space curve into a closed curve. The resulting simple loop (unknot U) or knot is denoted by K_a (recall that for sake of simplicity, the unknot U is also referred to as a knot). We shall analyze the resulting knot K_a on two levels:

a. by taking the corresponding Jones polynomial $V_{Ka}(t)$ of the knot K_a, and

b. by considering the projection of knot K_a to the original viewing plane, and finding new knots K_b which are compatible with the projection and preserve the most crossings; the Jones polynomials $V_{Kb}(t)$ of the new knots K_b are used to characterize the *projection*.

For most chain molecule problems, level a does not provide much information, since in most actual cases the simple unknot of the trivial polynomial $V_U(t) = 1$ is obtained. More information on the geometrical pattern of folding is retained on level b, that characterizes the given projection.

In fact, the characterization on level (b) is a special case of a more general characterization and reconstruction problem. If only a projection of a closed curve is given, and the crossing types are not specified, then one may ask the question: what is the family $\{K_b\}$ of knots compatible with a given projection, assuming no degenerate crossings? It is common that from experimental results such as images of knotted DNA fragments obtained by electron microscopy, only a single projection is available, where the crossing information is ambiguous, partially or fully missing. In these instances one faces a partial or the full reconstruction problem of knots from a given projection. Of course, the original knot K_a obtained in Steps 1 and 2 is sufficient to generate the given projection; however, there may be many more knots with the same projection. The family of all these knots gives a topological characterization of the projection.

On level b the task is to characterize the *projection,* without direct reference to the actual space curve K_a. By selecting one or several of the knots K_b that generate the same 2D projection (with crossing information supressed), and by using their Jones polynomials $V_{Kb}(t)$, a nonvisual, algorithmic characterization of the projection is obtained.

The actual projection to the viewing plane may well contain more crossings than the crossing number of the knot K_a of level a. Since the Jones polynomial $V_{Ka}(t)$ is independent of the actual number of crossings shown by the given projection, and it depends only on the identity of the knot K_a, the characterization of the knot K_a of level a by the Jones polynomial $V_{Ka}(t)$ does not provide a detailed enough characterization of the *projection* itself. By contrast, a more detailed characterization of the projection is obtained by the family of Jones polynomials $\{V_{Kb}(t)\}$ of the family $\{K_b\}$ of knots compatible with a given 2D projection (with crossing information supressed). In a somewhat simpler approach, a characterization of the projection is obtained using the polynomials of few, selected knots K_b from the family $\{K_b\}$.

In the general reconstruction problem we assume that the extension lines of Steps 1 and 2 of the conversion of the molecular space curve into the knot K_a add, at most, nondegenerate new crossings to the projection. This can always be achieved by an infinitesimal distortion of the molecular space curve model. The actual geometrical arrangement of the original knot K_a obtained in Steps 1 and 2 provides a description of the given projection. All n crossings of the projection can be characterized by the numbers $C_j = +1$ or -1, which are collected into a crossing vector

$$\mathbf{C} = (C_1, C_2, \ldots C_n). \tag{5.24}$$

By suitably modifying some or all n of these C_j numbers, all possible knots K_b with the same 2D projection (where the crossing information is suppressed) can be reconstructed, with arbitrarily chosen handedness for their crossings. By taking an n-dimensional *switching vector*

$$\mathbf{v} = (v_1, v_2, \ldots v_n), \tag{5.25}$$

with elements

$$v_i = +1, \text{ or } -1, \tag{5.26}$$

a new crossing vector $\mathbf{C}^v$ is generated from the reference crossing vector $\mathbf{C}$, by taking

$$\mathbf{C}^v = (C^v_1, C^v_2, \ldots C^v_n) \tag{5.27}$$

of elements

$$C^v_j = C_j v_j. \tag{5.28}$$

If crossing information for a reference projection is not available, then all elements of the reference crossing vector $\mathbf{C}$ may be chosen as unity. By taking all the 2^n possible n-dimensional vectors $\mathbf{v}$ of form (5.25), one obtains the crossing vectors $\mathbf{C}^v$ of all possible knots (and links) compatible with the given 2D projection (with crossing information supressed). The family of knots obtained is $\{K_b\}$, and the corresponding family of Jones polynomials is $\{V_{K_b}(t)\}$. Note that the same knot may be obtained by two or more different choices of vectors $\mathbf{v}$ and $\mathbf{C}^v$, and some choices of switching vectors $\mathbf{v}$ may be inconsistent with the 2D projection in the sense that they cannot lead to any knot.

In order to exploit the full characterization power of the Jones polynomials, it is of some interest to find those knots K^n_b of family $\{K_b\}$ that cannot have simpler 2D projections than the actual 2D projection of knot K_a. These are the knots K^n_b of family $\{K_b\}$ that have crossing numbers equal to n. If no such knot (or link) exists, then one may take K^n_b as a knot which has a crossing number that differs the least from the number of crossings in the projection.

Frequently, certain crossings of the projection cannot contribute to knottedness. These crossings can be eliminated from the knot model. We shall take n as the number of crossings obtained after eliminating those crossings that cannot contribute to knottedness. The actual Jones polynomials of these knots are in most

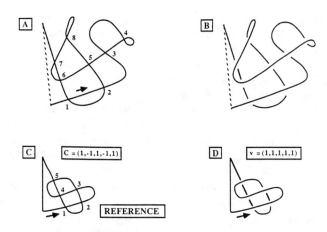

Figure 5.14 Four reference knots of the tertiary structure of the myoglobin molecule are shown. In part A of the figure, the oriented median line and its extension into knot is shown, where the endpoints are joined according to the rules described in the text. The crossings of the given planar projection are numbered consecutively from 1 to 8, whereas the actual crossing pattern is displayed in part B of the figure. For the construction of all possible knots compatible with the given planar projection shown in part A, the crossings 3, 4, and 8 are not essential and can be omitted. These crossings cannot contribute to knottedness in any of the possible knots compatible with the given planar projection. After eliminating these crossings, the reference projection of part C is obtained, where the remaining crossings are renumbered from 1 to 5. The actual crossing pattern can be specified by the crossing vector $C=(1,-1,1,-1,1)$, where the elements 1 and -1 represent right-handed and left-handed crossings, respectively. The elements of vector v indicate the switching of handedness relative to the reference vector C, where elements 1 and -1 indicate no switch and switch, respectively. The actual crossing pattern is displayed in part D of the figure, where the actual switching vector v, with reference to the reference projection of part C is also shown. The reference knot does not have to have the same handedness of crossings as those present in the actual 3D pattern; for example, if no 3D crossing information is available, then the reference vector C may be chosen with all its elements equal to 1.

instances different from, and more complicated than the polynomial of the knot K_a, hence they provide more detailed information on the projection.

The entire family of all the Jones polynomials $V_{K^n_b}(t)$ can be used for characterization. Alternatively, one may select just one of these polynomials according to the following criteria. The switching vectors v can be ordered by the lexicographic order (the order that would be used in a dictionary of n-letter words of an alphabet of just two letters, 1 and -1). This provides an ordering of the knots K_b, hence of knots K^n_b. One may choose the first K^n_b knot from the family $\{K_b\}$ for the characterization of the projection, and use its Jones polynomial $V_{K^n_b}(t)$ as a concise, nonvisual descriptor of the folding pattern.

In the example of Figures 5.14 and 5.15, a knot theoretical polynomial characterization of the folding pattern of myoglobin is given. In Figure 5.14, the folding of the myoglobin tertiary structure is represented by the projection A of knot B, generated following Steps 1 and 2. Among the eight crossings of the projection A, one can eliminate those of serial numbers 3, 4, and 8, since for no combination of the possible choices of crossings can they contribute to knottedness. The remaining crossings generate the reference projection C and the corresponding knot D. The reference crossing vector C of reference projection C is chosen as the actual crossing vector C=(1,-1,1,-1,1) of the reference knot D, however, in the general case the reference crossing vector C can be chosen arbitrarily. If no crossing information is available, then the vector C=(1,1,1,1,1) is chosen. For the actual choice of C=(1,-1,1,-1,1) there is no switch, and the switching vector v of reference knot D is v=(1,1,1,1,1).

In Figure 5.15, all knot types that are compatible with the reference projection C (where crossing information is suppressed) and their Jones polynomials are shown. The switching vectors v, given with respect to reference crossing vector C=(1,-1,1,-1,1), are also specified. In the figure the standard knot theoretical symbols, listed in Figure 3.5 are used. The "cake knot" denoted by $5_2^{\lozenge}$, where the symbol $\lozenge$ indicates that this knot is the mirror image of the standard "cake knot" 5_2, is the first knot in the lexicographic order of switching vectors v that has the maximum possible crossing number. In the case of the example this number is n=5. The corresponding Jones polynomial

$$V(t)= -t^{-6}+t^{-5}-t^{-4}+2t^{-3}-t^{-2}+t^{-1} \qquad\qquad (5.29)$$

is obtained from that of the standard "cake knot" 5_2, by following the rule for knot pairs that are mirror images, and replacing the variable t with t^{-1}. This polynomial provides a simple, nonvisual characterization of the given projection of the backbone of the myoglobin tertiary structure.

Figure 5.15 The collection of knots compatible with the planar reference projection of the myoglobin tertiary structure of Figure 5.13 and their Jones polynomials are shown. All the knots compatible with the projection can be constructed by generating all possible assignments of switching vectors v to the reference vector C. The ordering of switching vectors can follow the *lexicographic order* (i.e., the list of all five letter words in a dictionary where the alphabet has only two letters, 1 and -1, in this order). Most of the resulting knots are equivalent to the unknot U, whereas in the remaining cases the following knots are found: both trefoil knots 3_1 and $3_1^{\lozenge}$, the "figure 8 knot" 4_1, the "cake knot" 5_2, and its mirror image, $5_2^{\lozenge}$. Here the notations of Figure 3.5 are used. Besides the original reference unknot U of switching vector v=(1,1,1,1,1), only those knots are shown which are different from U and precede the unknot of vector v=(1,-1,-1,-1,-1) in the order provided for the vectors v. Beyond this vector in the lexicographic order, the remaining knots are the mirror images of those already obtained. The first occurrence of the highest possible crossing number has special significance: the corresponding knot is selected if no crossing information is available. For this projection, the highest crossing number 5, first occurs for the "cake knot" $5_2^{\lozenge}$. This knot is obtained using the switching vector v=(1,-1,1,-1,1). If only the projection of the backbone is given (i.e., if no crossing information is available), then the standard version 5_2 of this knot (i.e., not the mirror image $5_2^{\lozenge}$) and its Jones polynomial are selected for the topological characterization of the projection.

KNOT REPRESENTATIONS AND JONES POLYNOMIALS V(t) OF MYOGLOBIN TERTIARY STRUCTURE

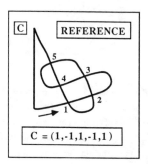

C | REFERENCE

$C = (1,-1,1,-1,1)$

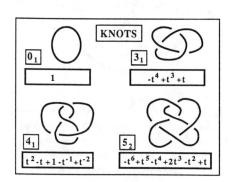

KNOTS

0_1

1

3_1

$-t^4+t^3+t$

4_1

$t^2-t+1-t^{-1}+t^{-2}$

5_2

$-t^6+t^5-t^4+2t^3-t^2+t$

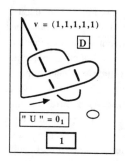

$v = (1,1,1,1,1)$

D

" U " $= 0_1$

1

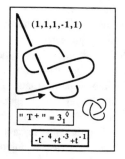

$(1,1,1,-1,1)$

" T+ " $= 3_1^{\Diamond}$

$-t^{-4}+t^{-3}+t^{-1}$

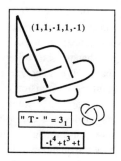

$(1,1,-1,1,-1)$

" T- " $= 3_1$

$-t^4+t^3+t$

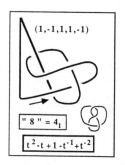

$(1,-1,1,1,-1)$

" 8 " $= 4_1$

$t^2-t+1-t^{-1}+t^{-2}$

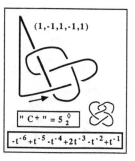

$(1,-1,1,-1,1)$

" C+ " $= 5_2^{\Diamond}$

$-t^{-6}+t^{-5}-t^{-4}+2t^{-3}-t^{-2}+t^{-1}$

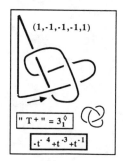

$(1,-1,-1,-1,1)$

" T+ " $= 3_1^{\Diamond}$

$-t^{-4}+t^{-3}+t^{-1}$

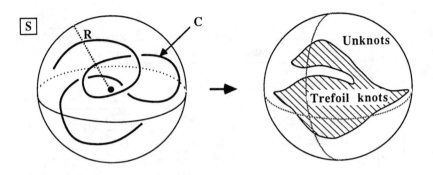

Figure 5.16 A direction-independent SGIM characterization of a space curve C, regarded as a molecular backbone. On the left-hand side the shape globe S of radius R is shown enclosing the space curve C. The centre of the sphere is chosen as the centre of mass of chain molecule C. On the right-hand side the shape invariance domains of the sphere are shown, as defined by the knot types derived from the projections. There are only two knot types in this example: unknots and trefoil knots.

In Figure 5.16, an illustration of the shape characterization of a molecular space curve by a knot polynomial SGIM is shown. The shape globe invariance domains are those derived from the projected knot patterns onto tangent planes of the shape globe S [112]. On the left-hand side of the figure a bounded molecular space curve is shown, surrounded by the shape globe S. On the right-hand side, the resulting subdivision of sphere into shape invariance regions is shown, with respect to the projected knots. Each of the regions is characterized by a different knot assigned to the curve C from the given viewing direction. In the particular case shown in the figure, the curve is rather simple, and the characterization requires only the unknot and the trefoil knot. The distribution of shape invariance regions on the shape globe S enclosing the molecular curve C provides a simple description of its shape. This approach eliminates the problems associated with the choice of a few arbitrary projections, since all the possible projection directions are taken into account.

CHAPTER

6

MOLECULAR SIMILARITY MEASURES AND MOLECULAR COMPLEMENTARITY MEASURES
THE QUANTIFICATION OF MOLECULAR SIMILARITY AND COMPLEMENTARITY

Similarity and complementarity are two fundamental aspects of all comparisons. When making comparisons, it is usual to search either for similarity or for complementarity. It is not a mere play on words that in most comparisons the merits of similarity and complementarity are similar: similarity and complementarity complement each other. The Latin saying "Similis simili gaudet", or "like likes like", as well as the saying "opposites attract" clearly apply in chemistry. For example, similarity is one of the guiding principles in solution chemistry ("like dissolves like"), whereas complementarity of shapes is important in many biochemical processes.

Molecular similarity is a concept often used, yet seldom clearly specified. If two objects are only similar but not equal, then the lack of equality to one another seems to invite the perception of some vagueness in their relation. This need not be so, but a precise definition of what is meant by the term similarity is seldom given. Notable early exceptions are various forms of the Hammond Postulate, interrelating stable species and transition structures occurring along reaction paths [352-360], and the precise wavefunction similarity measures of Carbó and co-workers [361-365] and Richards and co-workers [366,367].

It is only in recent years that systematic, comprehensive frameworks have been proposed for assessing molecular similarity and for quantifying the degree of similarity in chemistry (for critical expositions, see refs. [368,369]), using, in some instances, the concepts of topology [108,155-158,191,192,240,243,262,370]. The main difficulty lies in the complexity of molecules and their behavior: similarity may refer to one or another particular type of molecular property or process. Whether two molecules are judged similar or dissimilar is dependent on the context: the molecules of water and methane are similar in size, yet their chemical properties are very different. Even if one is concerned with a limited aspect of similarity such as molecular shape similarity, still some ambiguities prevail, since, depending on the context, different aspects of shape may be important. An assessment of similarity depends on the relative importance of these shape features.

6.1   Absolute and Relative Shape Analysis

In principle, absolute shape analysis methods do not rely on similarity arguments. For example, an (a,b) parameter map of a shape group analysis describes the shape of the given molecule without reference to any other similar or dissimilar molecule. Most of the shape analysis techniques described in Chapter 5 are applicable to single molecules, and a family of shape descriptors can be computed for each molecule. In their final form, the shape descriptors are usually given as numerical shape codes. These shape codes belong to the given molecule and they provide an *absolute shape characterization.* For n molecules, there are n shape codes of any given type and these codes can be stored in molecular data banks. If one is interested in assessing molecular shape similarity, a shape comparison between two molecules can be carried out at the level of the shape codes by considering these codes as vectors or matrices and comparing them numerically. The important advantage of absolute shape descriptors is that, once they are determined, there is no need to recompute them each time a molecule is compared to another. Shape similarity measures based on absolute shape descriptors are called *similarity measures of the first kind.*

By contrast, relative shape analysis and relative shape descriptors can change for each molecule, depending on the other molecule used for comparison. For n molecules there are $n(n-1)/2$ molecule pairs, hence $n(n-1)/2$ families of relative shape descriptors of the given type. Consequently, in the study of shape similarities in large molecular families, the quadratic dependence of the number of relative shape descriptors on the number of molecules is a disadvantage and the use of relative shape descriptors is often impractical. Shape similarity measures based on relative shape descriptors are called *similarity measures of the second kind.*

6.2   Visual, Computer Graphics Methods for Similarity Assessment by Inspection

The conventional approach of visual inspection is one of the simplest methods for similarity assessment. Using advanced computational chemistry and computer graphics techniques, three-dimensional images of molecular models, contour surfaces, or macromolecular representations can be displayed on the computer screen. For such computer images of molecular models, simple visual comparison can be used to judge molecular similarity. Visual comparisons are simple and are much enhanced by the chemical knowledge of the observer, who can take into account the known or assumed relative importance of various shape features seen on the computer screen. Nevertheless, such visual comparisons are often subjective and seldom reproducible. For example, while visually inspecting models of several hundred molecules during a long time interval, it is difficult to recall the details of a picture seen hours ago, in order to compare it to a current image. Furthermore, two different observers are likely to judge molecular similarity differently. Regarding the models of a thousand molecules, and trying to order them according to their similarity to a target molecule, two different observers are likely to order these molecules differently. These are potentially serious drawbacks of visual similarity search and assessment methods.

6.3   The Principles of Nonvisual, Algorithmic Similarity Analysis: Automated Similarity Assessment by Computer

In view of the subjective elements and the lack of reproducibility of visual inspection methods, it has appeared useful to develop computer techniques for evaluating the degree of similarity by reproducible algorithmic methods. By such nonvisual, computer-based algorithmic methods, similarity can be assessed and determined numerically. *Seeing is only believing, but computing is determining.*

Algorithmic determination, evaluation, and comparison of shapes are not simple problems. When dealing with shape, many of the relevant aspects are not easily representable numerically, as long as the full wealth of the detailed geometrical information is considered. This is the point where the power of topology becomes important, by focusing on the essential features and by describing them in terms of topological invariants. In fact, topological shape analysis methods incorporate many elements of visual inspection methods in a systematic and mathematically precise way. By quantifying shape and shape similarity topologically, the algorithmic computational methods of similarity assessment provide objectivity, reproducibility, and a justifiable degree of confidence in the results.

Consider a family of solid objects with well-defined boundaries. Their size can be characterized numerically. The topological shape group methods and related techniques are suitable for an algorithmic determination of their dominant shape features. However, this characterization becomes more complicated if the objects have no proper boundaries, for example, if fuzzy charge density clouds are compared. Nevertheless, the advantages of algorithmic techniques and automated similarity assessments by computer over visual inspection are evident in these cases too. By relying on isodensity contours, the shape similarity problem of the "boundaryless", fuzzy charge distributions can be solved by a repeated application of the shape analysis methods of continuous surfaces.

6.4   The GSTE Principle: Geometrical Similarity as Topological Equivalence

Several of the shape analysis methods described in the previous chapters had some common features:

1. infinitely many possible geometrical patterns and arrangements were classified by a combination of geometrical and topological criteria, and

2. the resulting classes were characterized by topological means.

For example, the changes in the nuclear geometry of a molecule are likely to alter the size, the location, and even the existence of shape domains (e.g., local curvature domains) on a MIDCO surface, but for most small changes of the nuclear geometry the existence and mutual neighbor relations of the shape domains remain invariant. There is a range of geometrical arrangements of the molecules with a common topological pattern of shape domains on their MIDCO surfaces, and the infinitely many geometrical arrangements within this range are regarded to belong to a single class. For this class of infinitely many different geometrical arrangements of the nuclei, the classification is based on a *geometrical condition* of certain bounds of the local curvature of the surface, and on the *topological condition* of having a certain pattern of neighbor relations of the various curvature domains. This combination of geometrical and topological conditions is in fact a shape condition, and the entire class of infinitely many nuclear arrangements satisfying this shape condition is characterized topologically by having a common family of shape groups. Within this class of nuclear arrangements the topological properties of the actual geometrical classification remain invariant, and all nuclear geometries of the molecule having the same topological relations among their MIDCO shape domains can be characterized by the same topological invariants as their shape descriptors. One can associate an abstract topological object with the entire class of geometrical arrangements, and characterize the class by the topological properties of the abstract object. The initial *geometrical classification* by curvature properties leads to an eventual *topological characterization.*

If the topological characterization gives the same results for two different nuclear arrangements of a given molecule then the two arrangements are *similar* in a geometrical sense. The same principle applies to two different molecules: if their topological shape characterizations give equivalent topological results, then the *two molecules are similar* in a geometrical sense. The *similarities* of the nuclear geometries within a range of the nuclear configurations of a given molecule, as well as the similarities in the shapes of two different molecules, are manifested in a *topological equivalence,* for example, in a topological equivalence of the shape domain patterns of a sequence of their MIDCO surfaces, resulting in a family of identical shape groups.

The above example illustrates a general principle of similarity analysis: *geometrical similarity corresponds to a topological equivalence* [108]. In order to formulate the above principle in more precise terms, we have to specify a general framework applicable to most molecular shape similarity problems. The (P,W)-*similarity* concept [108] provides such a general framework.

As we have seen in Chapters 2 and 4, there are various possibilities to select physical functions or molecular models for the *representation of molecular shapes,* and in Chapter 5 we have reviewed a variety of topological methods which can be applied and lead to *topological shape descriptors* for their characterization. When quantifying similarity of molecular shapes by topological techniques, it is necessary to specify the following [108]:

(i) the choice of *shape representation* P, taken as the physical property or model used to represent molecular shape [e.g., the complete 3D electronic charge density function $\rho(\mathbf{r})$, or a MIDCO surface G(a), or a MEPCO surface, or a fused sphere Van der Waals surface, or the backbone space curve of a folded protein],

(ii) the choice of the actual topological *shape descriptor* W, taken as the tool used for the characterization of P [e.g., the shape groups, or shape codes, or shape matrices of MIDCO's, or the distribution of Jones polynomials V(K) compatible with the folding pattern of a protein on a shape globe invariance map, SGIM].

This allows one to define the similarity of molecular shapes within the given (P,W) context in terms of a topological equivalence [108]. Informally, if we agree on which physical property or model P is to be used for comparison, and which topological features W of P are essential, then a topological equivalence of these features defines the similarity of the molecular shapes. Below we shall give a more formal definition.

The context within which similarity of molecular shapes is considered is defined by the choice of the pair (P,W) for a given property P and topological descriptor W. Consider two molecules, A and B. We can think of two abstract topological objects, denoted by $A_{(P,W)}$ and $B_{(P,W)}$, representing all the essential features of shape within the above (P,W) context and defined by the actual topological descriptor W for property P of molecules A and B, respectively. These abstract topological objects can be thought of as having no other properties except those specified and implied by the (P,W) pair. The shapes of two molecules A and B are (P,W)-*similar* [i.e., *similar within the context* (P,W)], denoted by

$$A \ (P,W) \ B, \tag{6.1}$$

if and only if the corresponding topological objects $A_{(P,W)}$ and $B_{(P,W)}$ are topologically equivalent:

$$A_{(P,W)} \ h \ B_{(P,W)}, \tag{6.2}$$

where h stands for the existence of a homeomorphic transformation h between the two objects [108].

Alternatively, if shape characterizations of molecules A and B are given by an actual (P_A, W_A) pair and a (P_B, W_B) pair, respectively, then

$$P_A = P_B, \tag{6.3}$$

and

$$W_A = W_B \tag{6.4}$$

imply that the two molecules A and B are (P,W)-similar,

$$A \ (P,W) \ B. \tag{6.5}$$

Similarity relation A (P,W) B is an equivalence relation. Clearly, any homeomorphism h realizing h in the definition (6.2) is reflexive, that is

$$A_{(P,W)} \ h \ A_{(P,W)}, \tag{6.6}$$

hence

$$A \ (P,W) \ A. \tag{6.7}$$

Also, h is symmetric, that is

$$\text{if } A_{(P,W)} \ h \ B_{(P,W)}, \text{ then } B_{(P,W)} \ h \ A_{(P,W)}, \tag{6.8}$$

hence

$$\text{if } A \ (P,W) \ B, \text{ then } B \ (P,W) \ A. \tag{6.9}$$

Furthermore, h is transitive, that is, for three molecules A, B, and C,

$$\text{if } A_{(P,W)} \ h \ B_{(P,W)}, \text{ and } B_{(P,W)} \ h \ C_{(P,W)}, \text{ then } A_{(P,W)} \ h \ C_{(P,W)}, \tag{6.10}$$

hence

$$A \ (P,W) \ B, \text{ and } B \ (P,W) \ C \text{ imply that } A \ (P,W) \ C, \tag{6.11}$$

that is, if A is (P,W)-similar to B, and if B is (P,W)-similar to C, then A is also (P,W)-similar to C.

The above general scheme [108] serves as the basis of constructing algorithms for nonvisual algebraic shape characterization. For a given (P,W) choice, a whole family of possible geometrical arrangements of different molecules may have a common actual realization of the shape descriptor W (e.g., all these arrangements may have a common shape group, or a common shape matrix). This implies that they can be represented by a common, abstract topological object $A_{(P,W)}$. Of course, for each family of arrangements, it is sufficient to find the topological invariants of the abstract object $A_{(P,W)}$. That is, all these arrangements can be collected into a family and the entire family can be represented by the given, common realization of the shape descriptor W, the actual shape descriptor W of the abstract topological object $A_{(P,W)}$. Each different realization of W is called a (P,W)-*shape type*, denoted by $\tau_{(P,W)}$, or simply by τ if the (P,W) pair is implied from context. For a given (P,W) pair, the various shape types τ_i are distinguished by some index i. With the exception of some degenerate cases, there are only a finite number of different shape types τ_i. Having a common shape type τ_i is in fact the same as having a (P,W)-similarity relation, that is, a *topological equivalence relation* representing a *geometrical similarity* of the chosen shape representations.

The above treatment of similarity is the basis of the *GSTE Principle:* treating *Geometrical Similarity as Topological Equivalence* [108].

The shape types τ_i are usually specified by various algebraic methods, for example, by a shape group or a shape matrix, or by some other algebraic or numerical means. The algebraic invariants or the elements of the matrices are *numbers,* and these numbers form a *shape code.* The (P,W)-shape similarity technique provides a nonvisual, algebraic, algorithmic shape description in terms of numerical shape codes, suitable for automatic, computer characterization and comparison of shapes and for the numerical evaluation of 3D shape similarity.

6.5   Whispered Messages and Similarity Sequences

One may note that our intuitive concept and actual evaluation of similarity used in everyday life do not necessarily follow the transitivity property (6.11) of the (P,W)-shape similarity relation. In a sequence of n objects, each object may appear similar to its immediate neighbors, but an observer may find the two objects at the two ends of the sequence as rather dissimilar. For example, the popular party game of lining up people and sending a whispered message through the line often produces a final message strikingly different from the initial one, to the amusement of the guests. Yet, assuming no deliberate mischief, the messages sent by two people who are neighbors in the line are likely to be similar. In some parts of Canada, the above game is known as "gossip". Indeed, true gossip also has the quality of gradual change as it is passed along, retaining some similarity between the stories heard and said by an individual, yet changing considerably throughout the grapevine. In such examples, the actual criteria we use to judge similarity may gradually change along the sequence of objects (in the actual example, along the sequence of messages) and the sequence may represent a whole range of similarities of varying nature.

Within the framework of (P,W)-shape similarity, the above problem can be treated by allowing the (P,W) pair to change gradually along the sequence of objects. We may require only that there exist *some* (P,W) pair that applies for each pair of objects that are neighbors in the sequence, and that the (P,W) pairs applied to pairs of objects not far from one another along the sequence are not too different. This approach conforms with our intuitive expectations and illustrates the generality and versatility of the (P,W)-shape similarity concept.

In the above problem, we require that *similar* (P,W) criteria, that is, *similar similarity criteria,* be used for pairs that are near to one another along the sequence. Hence, our task is to assess the similarity of the (P,W) criteria applied to the original objects. It is natural then to regard the similarity criteria as a set of new objects to be compared, and to use the very same method for assessing their similarity. The above scheme can be regarded as an *iterated similarity analysis,* since some similarity criterion is applied to the very similarity criteria (P,W) used for comparing the n original objects O_i, i = 1,2, . . . n,

The n-1, possibly different, (P,W) pairs applied to the n original objects O_i can be denoted by

$$(P^{(1)}_1, W^{(1)}_1), (P^{(1)}_2, W^{(1)}_2), \ldots (P^{(1)}_{n-1}, W^{(1)}_{n-1}). \tag{6.12}$$

These are the similarity criteria of level one within the iterative scheme, as indicated by the superscript (1). By regarding the above sequence of n-1 criteria *as a new set of objects,* a new higher-level similarity criterion [denoted as $(P^{(2)}, W^{(2)})$] is applicable to them. Of course, when judging the similarity of $(P^{(1)}_1, W^{(1)}_1)$ to $(P^{(1)}_2, W^{(1)}_2)$ and the similarity of $(P^{(1)}_2, W^{(1)}_2)$ to $(P^{(1)}_3, W^{(1)}_3)$, the similarity criterion [i.e., the $(P^{(2)}, W^{(2)})$ pair] does not have to remain constant, and one may, again, use different criteria. Consequently, further iterations are also possible, and the n-2 pairs of neighbors of the (n-1)-member sequence (6.12) can be judged by a sequence of n-2 similarity criteria

$$(P^{(2)}_1, W^{(2)}_1), (P^{(2)}_2, W^{(2)}_2), \ldots (P^{(2)}_{n-2}, W^{(2)}_{n-2}). \tag{6.13}$$

Here $(P^{(2)}_1, W^{(2)}_1)$ is used to compare $(P^{(1)}_1, W^{(1)}_1)$ to $(P^{(1)}_2, W^{(1)}_2)$, whereas $(P^{(2)}_2, W^{(2)}_2)$ is used to compare $(P^{(1)}_2, W^{(1)}_2)$ to $(P^{(1)}_3, W^{(1)}_3)$, and so on. The variations (in fact, the similarities) in the new sequence (6.13) of n-2 members can be judged by a further $(P^{(3)}, W^{(3)})$ similarity criterion, now applied to the n-3 neighboring pairs of the sequence of $(P^{(2)}_i, W^{(2)}_i)$ criteria. This $(P^{(3)}, W^{(3)})$ similarity criterion can also vary, resulting in yet another sequence

$$(P^{(3)}_1, W^{(3)}_1), (P^{(3)}_2, W^{(3)}_2), \ldots (P^{(3)}_{n-3}, W^{(3)}_{n-3}),$$

and possibly further sequences on subsequent levels, leading to the following iterative scheme:

$$
\begin{aligned}
&O_1, \qquad O_2, \qquad O_3, \qquad O_4, \qquad O_5, \ldots O_{n-1}, \qquad O_n \\
&(P^{(1)}_1, W^{(1)}_1), (P^{(1)}_2, W^{(1)}_2), (P^{(1)}_3, W^{(1)}_3), (P^{(1)}_4, W^{(1)}_4), \ldots (P^{(1)}_{n-1}, W^{(1)}_{n-1}) \\
&\quad (P^{(2)}_1, W^{(2)}_1), (P^{(2)}_2, W^{(2)}_2), (P^{(2)}_3, W^{(2)}_3), \ldots \ldots (P^{(2)}_{n-2}, W^{(2)}_{n-2}) \\
&\qquad (P^{(3)}_1, W^{(3)}_1), (P^{(3)}_2, W^{(3)}_2), \ldots \ldots (P^{(3)}_{n-3}, W^{(3)}_{n-3}) \\
&\qquad\qquad \ldots \ldots \\
&\qquad\qquad \ldots \ldots \\
&\quad (P^{(k-1)}_1, W^{(k-1)}_1), \ldots \ldots (P^{(k-1)}_{n-k+1}, W^{(k-1)}_{n-k+1}) \\
&\qquad (P^{(k)}_1, W^{(k)}_1). \ldots (P^{(k)}_{n-k}, W^{(k)}_{n-k}), \\
&\qquad\qquad \ldots \ldots
\end{aligned}
\tag{6.14}
$$

Here the original objects, denoted by O_i, $i = 1, 2, \ldots n$, can be regarded as the sequence at level zero, $j=0$. In general, on the j-th level ($j>0$), there are at most $n-j$ different criteria $(P^{(j)}_i, W^{(j)}_i)$, $1 \le i \le n-j$, and the criterion $(P^{(j)}_i, W^{(j)}_i)$ is used to compare the neighboring members $(P^{(j-1)}_i, W^{(j-1)}_i)$ and $(P^{(j-1)}_{i+1}, W^{(j-1)}_{i+1})$ of the $(n-j+1)$-member sequence on the previous level $j-1$.

The first level k where a *common* criterion

$$(P^{(k)}, W^{(k)}) = (P^{(k)}_1, W^{(k)}_1) = \ldots = (P^{(k)}_i, W^{(k)}_i) = \ldots = (P^{(k)}_{n-k}, W^{(k)}_{n-k})$$

is already applicable for the entire previous sequence

$$(P^{(k-1)}_1, W^{(k-1)}_1), (P^{(k-1)}_2, W^{(k-1)}_2), \ldots (P^{(k-1)}_{n-k}, W^{(k-1)}_{n-k})$$

at level k-1 is analogous to the *factorial level* of difference sequences of powers of integers.

In order to use this analogy, we describe a result of some interest in number theory. In general, for the sequence

$$d^{(0)}_0 = 0^k, \ d^{(0)}_1 = 1^k, \ d^{(0)}_2 = 2^k, \ d^{(0)}_3 = 3^k, \ d^{(0)}_4 = 4^k, \ d^{(0)}_5 = 5^k, \ \ldots \tag{6.15}$$

of the k-th powers of integers, the k-th difference sequence turns out to be the constant k! These difference sequences are defined iteratively, for example, the ith element $d^{(j)}_i$ of the jth sequence is

$$d^{(j)}_i = d^{(j-1)}_{i+1} - d^{(j-1)}_i. \tag{6.16}$$

For example, the sequence 0, 1, 4, 9, 16, 25, ... of squares of integers 0, 1, 2, 3, 4, 5, ... has the first difference sequence 1, 3, 5, 7, 9, ..., and the second difference sequence (the sequence of differences between subsequent elements of the first difference sequence) is 2, 2, 2, 2, ... That is, for the second powers the second difference sequence is constant, and is equal to 2!=2. In general, in the k-th difference sequence one obtains a constant, the factorial k! of the exponent k of the k-th powers of integers. For example, for the seventh powers the seventh difference sequence contains the constant 7! In the scheme below, the case of the fourth powers is shown in detail:

	0^4,	1^4,	2^4,	3^4,	4^4,	5^4,	6^4,	7^4,	8^4,	9^4,	10^4, ...
	0,	1,	16,	81,	256,	625,	1296,	2401,	4096,	6561,	10000, ..
(k=1)		1,	15,	65,	175,	369,	671,	1105,	1695,	2465,	3439, ...
(k=2)			14,	50,	110,	194,	302,	434,	590,	770,	974,	1202 ...
(k=3)				36,	60,	84,	108,	132,	156,	180,	204,	228, ...
(k=4)					24,	24,	24,	24,	24,	24,	24,	24,	24, ...

$$\tag{6.17}$$

In the fourth difference sequence (k=4) the constant difference 4!=24 is obtained.

By virtue of this analogy, the serial number k of the first level where a common criterion $(P^{(k)}, W^{(k)})$ is already applicable is called the *power of the similarity sequence,* and the final condition $(P^{(k)}, W^{(k)})$ is called the *factorial similarity condition.*

If the two molecules A and B turn out to be dissimilar by a given (P,W)-shape similarity criterion [i.e., if they do not fulfill the equivalence relation A (P,W) B], then the differences between their numerical shape descriptors can serve as a *dissimilarity measure.* That is, for a (P,W)-dissimilar molecule pair A and B, the (P,W)-similarity concept allows one to quantify *how* different their topological invariants are. A simple and straightforward approach is based on a simple vector comparison of the lists of Betti numbers of the shape group technique, or on the numerical comparison of shape matrices.

The technique of shape globe invariance maps (SGIM) leads to a particular realization of the (P,W)-similarity concept. The actual shape representation P, for example, an isodensity contour with relative local convexity domains or a space curve representing a backbone of a chain molecule, is enclosed within a shape globe S. The topological characterization is given by a shape descriptor W, where W is the shape globe S, together with a topological descriptor of the pattern of spherical domains of a selected invariant. For example, the shape descriptor can be chosen as the shape matrix of the pattern projected to tangent planes of S, or as the knots derived from the crossing pattern of chain molecule images projected to the tangent planes of the shape globe. The final result of such analysis is a numerical shape code that can be compared by algorithmic methods to those of other molecules.

6.6   The Fundamentals of Resolution Based Similarity Measures (RBSM)

The principle of the resolution based similarity measures is illustrated in Figure 6.1. Consider three objects, A, B, and C, of comparable sizes, observed from a great distance. For a distant observer, all three objects appear as mere points, and the objects cannot be distinguished. If the objects are somewhat closer, then A may already show some distinctive shape feature, yet B and C may still appear indistinguishable. In this case the observer must approach B and C much closer in order to distinguish them. We may conclude that B and C are more similar to each other than A is to B or A is to C, simply, because it took a closer look to distinguish B from C. Alternatively, the observer may use a series of binoculars, and in order to distinguish the objects, a higher level of resolution of the observed picture is required if the objects are more similar. One can define a *similarity measure* based on the level of resolution required to distinguish objects. This conclusion and the above example illustrate the principle of *Resolution Based Similarity Measures* (RBSM's), described in detail in [240] and [243].

The concept of resolution can be approached within a general topological framework. Consider a family of objects and a hierarchy of topologies defined for each object, where the hierarchy is ordered by the finer-cruder relations of the topologies (see Section 3.2). Considering finer topologies is analogous to considering higher levels of resolution. For example, take a family of MIDCO

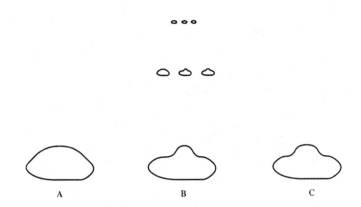

Figure   6.1 Illustration of the principle of resolution based similarity measures (RBSM). Three objects, A, B, and C, appear indistinguishable at a great distance (low resolution). At some closer distance (medium resolution), object C is already distinguishable from A and B, but the latter two still appear indistinguishable. At a close distance (high resolution) A and B are also distinguishable. A numerical similarity measure can be defined in terms of the resolution required to distinguish objects: if two objects are very dissimilar, they are distinguishable at a low resolution; if they are more similar, then a higher resolution is required to distinguish them; and if the shapes of the two objects are identical, then they are indistinguishable even at infinite resolution.

surfaces, and several shape domain partitionings for each surface. For each shape domain partitioning, the set of shape domains can be regarded as a defining subbase for a topology on the MIDCO, turning the MIDCO into a topological space. If the shape domains of a cruder partitioning can be constructed as unions of the shape domains of a finer partitioning, then the corresponding topologies are also related by a cruder-finer relation, as described in Chapter 3. If all partitionings can be ordered by such relations, then the corresponding hierarchy of topologies also provides a hierarchy of resolutions. Two MIDCO's with shape groups identical at a finer shape domain partitioning are more similar than two MIDCO's which have different shape groups already at a cruder shape domain partitioning. The complexity of the partitioning, that is, a measure of how fine is the corresponding topology, gives a *resolution based measure of similarity* of MIDCO surfaces.

6.7   Molecular Similarity Measures and Chirality Measures Based on Resolution and Fuzzy Set Theory

The first example of similarity measure we shall consider is a *resolution based similarity measure* (RBSM). This particular realization of a RBSM is conceptually simple, but it is not recommended for highly detailed shape comparisons since its practical applications are computationally feasible only for relatively low levels of resolution [240,243].

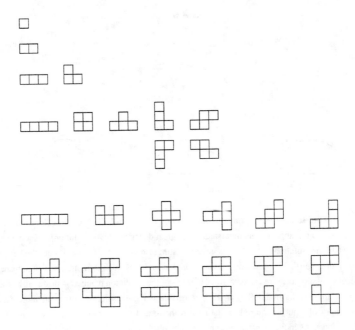

Figure 6.2 Lattice animals of less than six cells. The mirror images of chiral animals are given in pairs, with respect to horizontal reflection planes. The animals of four cells are "Skinny", "Fatty", "Knobby", "Elly", and "Tippy", in the order listed in the figure.

For simplicity, we shall first consider the shapes and similarities of a finite number of different planar domains. The level of resolution of observing these domains can be quantified by placing these domains on square grids of various sizes in the plane, and observing which squares of the grid fall within the planar domains. The pattern of these squares gives a discretized approximation to the shape of the planar domain, and the grid size can be considered as a measure of resolution. Such square grid patterns, if connected by edge contacts between the squares, are called *square cell configurations* or *lattice animals*. The lattice animals which contain no more than five squares of the underlying square lattice are listed in Figure 6.2. Such lattice animals have interesting mathematical properties and provide simple models for a great variety of chemical problems; for example, for modeling the patterns of adsorbed molecules on metallic surfaces, and for the description of percolation problems [54,240,243,371-380]. Note that, in the present context, animals that are chiral mirror images of each other are considered different. For each planar domain to be characterized, the initial grid size can be chosen large enough so that no square fits within the domain, but a small decrease of this grid size permits one or more squares to fit within. In Figure 6.3 the relations of three planar domains and various lattice animals enclosed by them are considered. For each of these three domains, the case of a single cell, n=1, corresponds to the initial

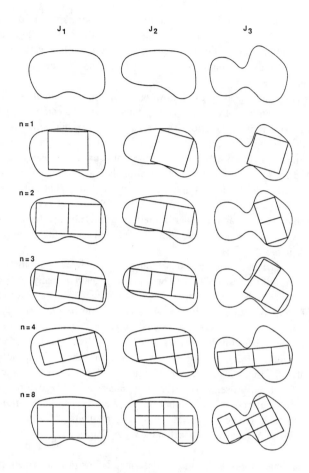

Figure 6.3 Some of the interior filling animals of three Jordan curves, $J_1, J_2,$ and J_3, enclosing three planar domains, $D_1, D_2,$ and D_3, of different shapes. For cell numbers 1 and 2 the the resolution is not sufficient to distinguish these curves. For cell number n=3 and for any higher cell number, the third curve, J_3, has interior filling animals different from those of curves J_1 and J_2, but curves J_1 and J_2 are distinguishable only for cell number n=8 and beyond. Accordingly, the greatest degree of similarity is found between curves J_1 and J_2 (and the respective domains D_1 and D_2), in agreement with expectation based on visual inspection.

grid size. As shown by the example, this initial grid size can be different for different domains. In order to avoid dealing with domain-dependent grid sizes, the resolution can be defined relative to each domain, and can be characterized by the number of squares ("cells") occurring in the animals associated with the domain. Clearly, scaling the domain as well as the animal inscribed within the domain by a factor of two will not change their mutual relations. The level of resolution will be judged by the number of square cells which fit within the domain and not by the actual grid size directly; this allows one to compare the shapes of domains of very different sizes. By gradually increasing the number of cells, higher and higher levels of resolutions can be obtained, where the resolution is always relative to the size of the individual domain. At higher resolution, fine details of shape and similarity can be recognized.

A fit is characterized by the following condition: an animal A fits a planar domain D if no animal of the same cell size and more cells can be inscribed within D. Starting with the initial grid size for each domain, and simultaneously and gradually decreasing these grid sizes (i.e., increasing the number of cells), one can monitor which lattice animals fit within the domains. In the early stages of this process, typically, many common lattice animals occur for all the domains. For example, in the first stage the single cell animal is likely (but not necessary) to appear for each domain. However, for a greater number of cells, a given lattice animal is likely to appear for fewer domains, and eventually, for a large enough number of cells, no common lattice animals are found within the given finite family of different planar domains.

Consider just two different planar domains. For low resolutions (i.e., for grid sizes allowing only a few cells to fit within the domains), many of the inscribed animals are likely to be common for the two planar domains. However, the number of cells of animals fitting within the two domains can be increased by gradually decreasing these two grid sizes. A special stage in this process is of particular importance: the stage with the smallest number of cells, such that for this and for any larger number of cells all the lattice animals inscribed within the two planar domains are different. The number of squares of the corresponding inscribed animals at this stage can serve as a *grade of similarity* of the two planar domains (in a formal treatment, this number is referred to as the *similarity index*). Evidently, no such number can be found for two domains of identical shapes (but possibly of different sizes), hence for identical shapes the grade of similarity is infinite.

An implementation of this idea is described below in more detail, providing a size-independent scale for the level of resolution [240,243]. We shall use the concept of Jordan curves. A Jordan curve J divides the plane into two parts: a bounded domain (the interior of the Jordan curve) and the remaining, unbounded subset of the plane (the exterior of the Jordan curve). For example, in Figure 6.3, the boundaries of planar domains are such Jordan curves, denoted by J_1, J_2, and J_3, respectively. Consider a Jordan curve J in the plane. The interior D of J is the planar domain we wish to characterize at various levels of resolution. Domain D can be modeled by square-cell configurations; for example, one may place J on a square grid and consider the family of all square cells falling within the interior of J. Subject to the constraint of edge-connectedness of cells and the requirement that the planar set covered by the cells considered must be simply connected, the corresponding square-cell configurations are called lattice animals (or in short, animals). Two animals which can be converted into each other by scaling, translation and rotation, are considered the same; consequently, animals provide natural, size-independent tools to determine the level of resolution of various shape descriptions. Since we shall use the term "animal" only for simply connected square cell configurations, all configurations with "holes" or separate, disconnected parts are excluded.

The smaller the squares of the grid, the better the resolution of the representation of D by the animals. By approximately filling up the interior D of J by animals at various levels of resolution, a shape characterization of the continuous Jordan curve J can be obtained by the shape characterization of animals. The animals contain a finite number of square cells, consequently, their shape characterization can be accomplished using the methods of discrete mathematics. As a result, one obtains an approximate, *discrete* characterization of the shape of the Jordan curve (i.e., the shape of a *continuum*). The level of resolution can be represented indirectly, by the number of cells of the animals. In particular, one can show [240,243] that the number of cells required to distinguish between two Jordan curves provides a numerical measure of their similarity.

If one chooses a small enough size s for the length of the side of the square cells, then any finite animal can fit within the given planar domain D. Whether an animal A fits within the interior D of a given Jordan curve J depends on the relative size of J and the cells of the animal. For a given Jordan curve J and a given cell size s there exists a countable family $F(J,s)$ of animals which fit within domain D. Clearly, if the size s is too large, then this family is empty. With reference to J and s, the members $A_i(J,s)$ of this family $F(J,s)$ are called the *inscribed animals* of D [240,243].

For a given J and s pair there exists a maximum number n of cells for inscribed animals. A smaller cell size s is associated with the same or a larger number of cells and with a better resolution, hence with a better approximation. However, a small change of cell size s does not necessarily change n. For this reason, it is advantageous to refer directly to n instead of s. The n-cell animal $A_i(J,n)$ inscribed in the interior D of Jordan curve J is an *interior filling animal* of J if and only if no animal of the same cell size s and more than n cells can be inscribed in J [240,243]. In particular, none of the interior filling animals $A_i(J,n)$ can be enlarged by a cell and still fit within the interior D of J.

With respect to size, the actual level n of resolution depends on the relative size of D as compared to size s of the cells of the animals. This relative size is implied by the maximum number n of cells which fit within domain D, and in the $A_i(J,n)$ notation the cell size information s is not given directly.

The family $F(J,n)$ of all interior filling animals $A_i(J,n)$ of the given Jordan curve J,

$$F(J,n) = \{A_i(J,n)\}, \tag{6.18}$$

provides an absolute shape characterization of J and its interior D at the level n of resolution. These $F(J,n)$ sets are also suitable to introduce a relative measure for shape similarity of two Jordan curves J_1 and J_2.

In Figure 6.3, three Jordan curves, J_1, J_2, and J_3, are shown, with some of their interior filling animals. At both levels n=1 and n=2 there is only one interior filling animal, common to all three curves. Hence, at these levels of resolution the shapes of J_1, J_2, and J_3 appear the same. At level n=3, however, only the two curves J_1 and J_2 have the given interior filling animal in common; it is different from the animal shown for the curve J_3. At this and at all higher levels of resolution, all interior filling animals of curve J_3 are different from those of curves J_1 and J_2. However, one must reach the level n=8 in order to find that all interior filling animals of curves J_1 and J_2 are different. For the actual curves, J_1 and J_2, this difference prevails at all higher levels. Since it requires a higher level of resolution (n=8) to distinguish the shapes of the pair J_1 and J_2 than the level (n=3) needed for either of the pairs J_1 and J_3 or J_2 and J_3, we conclude that the closest similarity is between the shapes of J_1 and J_2. Most observers would find the same conclusion based on visual inspection.

The above example illustrates the motivation for the choice of a *similarity index* $i_0(J_1,J_2)$ of two Jordan curves J_1 and J_2, defined as the smallest n_c value at and above which all interior filling animals of Jordan curves J_1 and J_2 are different, that is,

$$i_0(J_1,J_2) = \begin{cases} \min \{n_c: F(J_1,n) \cap F(J_2,n) = \emptyset \text{ if } n \geq n_c\}, & \text{if such minimum} \\ & \text{exists,} \qquad (6.19) \\ \infty \quad \text{otherwise.} \end{cases}$$

If the shapes of the two domains enclosed by the Jordan curves J_1 and J_2 are identical (i.e., if they can be obtained from one another by scaling), then no finite n_c value exists and the similarity index $i_0(J_1,J_2) = \infty$. For curves J_1 and J_2 of different shapes, the more similar their shapes, the greater the cell number n of the largest common interior filling animals. Consequently, the similarity index $i_0(J_1,J_2)$ is a large number if the two Jordan curves J_1 and J_2 are very similar, and $i_0(J_1,J_2)$ is a small number for highly dissimilar curves.

The *degree of dissimilarity* $d(J_1,J_2)$ is defined in terms of similarity index $i_0(J_1,J_2)$ as follows:

$$d(J_1,J_2) = 1 /(i_0(J_1,J_2) - 2). \qquad (6.20)$$

The smallest cell number n at which there exist different animals is three, hence the smallest possible value for $i_0(J_1,J_2)$ is also 3; that justifies the inclusion of the number 2 in the denominator. As a consequence of this definition, the degree of dissimilarity $d(J_1,J_2)$ may take values from the [0,1] interval, where greater values indicate greater dissimilarity. For two Jordan curves J_1 and J_2 of identical shapes $d(J_1,J_2) = 0$.

The *degree of similarity* $s(J_1,J_2)$ is defined in terms of the degree of dissimilarity $d(J_1,J_2)$ as

$$s(J_1,J_2) = 1 - d(J_1,J_2). \qquad (6.21)$$

If two Jordan curves J_1 and J_2 have identical shapes, then their degree of similarity $s(J_1,J_2) = 1$, otherwise $s(J_1,J_2)$ is a smaller positive number.

The similarity index $i_0(J_1,J_2)$ and the degree of similarity $s(J_1,J_2)$ of two Jordan curves J_1 and J_2 are of general applicability for the evaluation of the

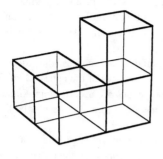
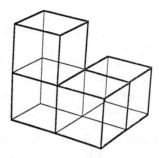

Figure 6.4 Mirror image pair of the smallest chiral polycubes.

similarities of planar domains D_1 and D_2, regarded as the interiors of these Jordan curves. In chemistry, these tools for the quantification of similarity are applicable in a wide variety of fields, for example, for the shape comparison of curves defined as cross-sections of molecular contour surfaces, or contours of molecular aggregates, or patterns of molecules adsorbed on metallic surfaces, important in studies of catalysis.

The main ideas of the above shape characterization technique and the concept of the degree of similarity have been extended to three-dimensional objects such as formal molecular bodies and molecular boundary surfaces [240,243]. The actual tools for this purpose are *polycubes* which are the three-dimensional analogues of square-cell configurations [240,243].

A connected arrangement of a finite number n of impenetrable cubes C of uniform edge length s is called a polycube, if only three types of contacts between cubes are allowed: common face, common edge, and common vertex. If $n>1$, then each cube of the polycube P must have a face contact with another cube of P. One may regard polycubes as parts of a cubic lattice. In Figure 6.4, the pair of smallest chiral polycubes is shown. These polycubes contain four cubes.

The polycubes P we consider for similarity analysis fulfill the following three restrictions:

1. if there is an edge contact between two cubes C and C' of P then there must also be a face contact between C and C', or there must exist a cube C'' having face contact with both C and C',

2. if there is a vertex contact between two cubes C and C' of P then there must also be either an edge contact between them, or there must exist two cubes C'' and C''' with face contact to each other and cube C'' having face contact to C and cube C''' having face contact to C' (see illustration in Figure 6.4),

3. the polycube P, regarded as a single body, is topologically equivalent (according to the metric topology of the 3D space) to the 3D body it represents. In the most common case, a formal molecular body $B(a)$ is topologically equivalent to a solid ball, however, toroidal or other, more complicated topologies are also possible.

Conditions 1. and 2. ensure that the polycubes are facewise connected, whereas condition 3. is the natural requirement that the three-dimensional molecular body and its representations at various levels of resolution are not incomparably different.

By analogy with the perimeters of animals and Jordan curves, the *surface* G(P) of a polycube P is the point set union of all those faces of the cubes C of P that are on precisely one cube. The surfaces of polycubes are used to approximate MIDCO surfaces G(a), and to characterize the shapes of the formal molecular bodies B(a) enclosed by them.

The size of the cubes C is characterized by the uniform edge length s. By gradually decreasing s and increasing the number n of cubes in the polycubes P inscribed within G(a), one can approximate the formal molecular body B(a) at increasing levels of resolution.

Consider a given molecular contour surface G(a). If the size s of the cubes is chosen small enough, then any finite polycube P can fit within G(a). As in the two-dimensional case, we do not consider orientation constraints and we assume that the contour surface G(a) and polycube P may be translated and rotated with respect to one another; the relative orientation of G(a) and the cubic grid is not fixed. In this model, the identity of a polycube is independent of its orientation. Two polycubes P and P' are regarded identical if and only if they can be superimposed on one another by translation and rotation in 3D space. Note, however, that the polycube method of shape analysis and determination of resolution based similarity measures can be augmented with orientation constraints, suitable for the study of molecular recognition and shape problems in external fields or within enzyme cavities [240,243].

A polycube $P_i(G(a),n)$ of n cubes is an *interior filling polycube* of the contour surface G(a) if and only if no polycube P of the same cube size s and of n+1 cubes can be inscribed in G(a).

The three-dimensional RBSM method relies on the shape properties of interior filling polycubes $P_i(G(a),n)$ inscribed in molecular contour surfaces G(a) when assessing the similarity of the G(a) contours and the formal molecular bodies B(a) enclosed by them. In order to define levels of resolution scaled relative to the molecular size, the absolute size parameter s is not used directly. One obtains more comparable shape characterizations of both small and large objects when using the *same number* of cubes. Consequently, each level of resolution is defined by the number n of cubes of interior filling polycubes $P_i(G(a),n)$, which depends on the relative size of the object G(a) as compared to the cube size s.

The family of all interior filling polycubes $P_i(G,n)$ of the molecular contour surface G at level n is denoted by F(G,n),

$$F(G,n) = \{P_i(G,n)\}. \tag{6.22}$$

This set F(G,n) provides an absolute shape characterization of G and the body B enclosed by it. By analogy with the two-dimensional case, we may use these F(G,n) sets to introduce a relative measure for shape similarity of two molecular contour surfaces G_1 and G_2. These surfaces may belong to two different molecules, or they can be two MIDCO's of the same molecule with two different contour density values a_1 and a_2.

The *similarity index* $i_0(G_1,G_2)$ of two molecular contour surfaces G_1 and G_2 is the smallest n_c value at and above which all interior filling polycubes of contour surfaces G_1 and G_2 are different,

$$i_0(G_1,G_2) = \begin{cases} \min\ \{n_c: F(G_1,n) \cap F(G_2,n) = \emptyset \text{ if } n \geq n_c\}, & \text{if the minimum} \\ & \text{exists} \qquad (6.23) \\ \infty \quad \text{otherwise.} \end{cases}$$

If, for two contour surfaces G_1 and G_2 of nonidentical shapes, the cell number n of a largest common interior filling polycube is a large number, then we perceive these surfaces as similar. This perception is reflected in the similarity index $i_0(G_1,G_2)$, which is a large integer number if the two contour surfaces G_1 and G_2 are very similar, and a smaller integer value if G_1 and G_2 are highly dissimilar. If two molecular contour surfaces, G_1 and G_2 can be obtained from one another by scaling, then their shapes are identical. For contour surfaces G_1 and G_2 of identical shapes no finite n_c value exists, consequently, their similarity index $i_0(G_1,G_2) = \infty$.

The *degree of dissimilarity* $d(G_1,G_2)$ has been defined [240,243] as follows:

$$d(G_1,G_2) = 1 /(i_0(G_1,G_2)\text{-}2). \qquad (6.24)$$

The smallest cube number n at which there exist different polycubes, hence the smallest possible similarity index $i_0(G_1,G_2)$ is 3; this is reflected by the inclusion of the number 2 in the denominator. Note that the same number appears in the formula for the two-dimensional degree of dissimilarity. The degree of dissimilarity $d(G_1,G_2)$ takes values from the $[0,1]$ interval, greater values indicating greater dissimilarity. Evidently, if the contour surfaces G_1 and G_2 have *identical shapes* then their degree of dissimilarity is zero, $d(G_1,G_2) = 0$, even if the *sizes* of G_1 and G_2 are different.

The *degree of similarity* $s(G_1,G_2)$ of two molecular contour surfaces G_1 and G_2 is defined as

$$s(G_1,G_2) = 1 - d(G_1,G_2). \qquad (6.25)$$

If the two molecular contour surfaces G_1 and G_2 have identical shapes then their degree of similarity $s(G_1,G_2) = 1$, otherwise it is a smaller positive number.

The quantification of chirality has been discussed in Chapter 1. A family of intuitively appealing chirality measures is based on the maximum overlap between a chiral object and its mirror image, as described by Kitaigorodskii [46], Gilat and Schulman [48-50], and Mislow, Buda, and Auf der Heyde [52,53,57,58]. Alternative approaches have been proposed using reference objects and the Haussdorf distance of point sets for characterization by Rassat [47], or using fuzzy set representations [381] of chirality with reference to fuzziness in an epistemological sense by Mislow and Bickart [27], or using the principle of energy-weighted fuzzy achirality resemblance of Mezey [55], based on the syntopy model developed by Mezey and Maruani [252]. For more detail on the subject the reader should consult references [46-57], and the review by Mislow and coworkers [58].

Here we shall consider a quantification technique of chirality based on the degree of resolution. In the context of RBSM's, one should note that the detection of the presence or the lack of chirality is also resolution-dependent. Although chirality is an absolute property, its detection is not, and this resolution-dependence allows one to introduce a formal scale for chirality. If chirality is already detectable at a low level of resolution, then it is justified to regard the object "more chiral" than another object that reveals its chirality only at a much higher level of resolution.

For both two- and three-dimensional chirality, a formal degree of chirality has been introduced [54,55], based on a discretization of shape features using lattice animals and polycubes [240,243]. These definitions are based on chiral animals and chiral polycubes, for which chirality can be detected by simple algebraic means.

An animal A is achiral if and only if A can be superimposed on its mirror image $A^\Diamond$ by translation and rotation within the plane:

$$A = A^\Diamond .\tag{6.26}$$

Otherwise the animal A is chiral.

We say that Jordan curve J is chiral at and above cell number n_χ if each interior filling animal $A_i(J,n)$ is chiral if $n \geq n_\chi$. The *chirality index* $n_\chi(J)$ is the smallest cell number $n_\chi = n_\chi(J)$ at and above which all interior filling animals $A_i(J,n)$ are chiral,

$$n_\chi(J) = \begin{cases} \min\{ n_\chi : \text{each } A_i(J,n) \text{ is chiral if } n \geq n_\chi\}, & \text{if the minimum exists} \\ \infty & \text{otherwise.} \end{cases}\tag{6.27}$$

Since the smallest chiral lattice animals have four cells [54], the minimum possible value for chirality index is $n_\chi(J)=4$. The *degree of chirality* $\chi(J)$ of a Jordan curve J is defined as

$$\chi(J) = 1 / (n_\chi(J)-3).\tag{6.28}$$

This measure of chirality gives the value 1 for "very chiral" curves and 0 for achiral ones.

A similar treatment has been applied for the three-dimensional case [55], using polycubes. A polycube P_n is achiral if and only if P_n can be superimposed on its mirror image $P^\Diamond_n$ by translation and rotation:

$$P_n = P^\Diamond_n .\tag{6.29}$$

Otherwise the polycube P_n is chiral.

A molecular contour surface, such as a MIDCO surface $G(a)$ is chiral at and above cube number n_χ if each interior filling polycube $P_n(G(a))$ of $G(a)$ is chiral if $n \geq n_\chi$. The *chirality index* $n_\chi(G(a))$ is the smallest n_χ value at and above which all interior filling polycubes $P_n(G(a))$ are chiral:

$$n_\chi(G(a)) = \begin{cases} \min\{n_\chi : \text{each } P_n(G(a)) \text{ is chiral if } n \geq n_\chi\}, \text{ if the minimum} \\ \qquad\qquad\qquad\qquad\qquad\qquad\qquad\qquad\qquad\qquad \text{exists} \\ \infty \quad \text{otherwise.} \end{cases} \qquad (6.30)$$

The degree of chirality $\chi(G(a))$ of a molecular surface $G(a)$ is

$$\chi(G(a)) = 1 / (n_\chi(G(a))-3). \qquad (6.31)$$

The smallest chiral polycube has four cubes (as shown in Figure 6.4), and the number 3 in the denominator ensures that the degree of chirality takes values from the [0,1] interval.

The principle of the RBSM method is also applicable for a more direct quantification of chirality. The degree of similarity between an object and its mirror image can be used as a simple measure of chirality. In particular, such a measure can be based on the degree of similarity $s(G,G^\Diamond)$ or $s(J,J^\Diamond)$ between the object G or J and its mirror image $G^\Diamond$ or $J^\Diamond$, in the three- or two-dimensional cases, respectively.

The *similarity based measures* $\alpha_S(G)$ and $\alpha_S(J)$ *of achirality* are defined as

$$\alpha_S(G) = 2 \, s(G,G^\Diamond) - 1 \qquad (6.32)$$

and

$$\alpha_S(J) = 2 \, s(J,J^\Diamond) - 1, \qquad (6.33)$$

respectively.

At very low resolutions most chiral objects appear achiral. The measures $\alpha_S(G)$ and $\alpha_S(J)$ refer to the lowest resolutions at which the objects already appear chiral, as measured by their similarity to their mirror images. Note that in the general case the smallest possible value of both the two- and three-dimensional similarity indices is 3. However, for mirror images, the smallest possible value for both similarity indices is 4. This is the value obtained for the enantiomeric pair of one of the smallest chiral animals, (called "Tippy", see, *e.g.* ref. [54]), and also for the enantiomeric pair of the smallest chiral polycubes, the four-cube screws shown in Figure 6.4, where the four cubes of the polycube are arranged as in the "vertex contact" condition 2. discussed above. Since a similarity index value of 3 is unattainable, a rescaling of the $s(G,G^\Diamond)$ and $s(J,J^\Diamond)$ measures is required, as given in Equations (6.32) and (6.33). The coefficient 2 and the term -1 in the above expressions ensure that these achirality measures are taking values from the [0,1] interval. Objects which are "fully achiral", that is, objects which appear achiral at any level $n \geq 4$ of resolution, have an achirality measure of $\alpha_S = 1$, whereas objects showing the greatest dissimilarity with their mirror images have an achirality measure of $\alpha_S = 0$.

Using the above achirality measures, the *similarity based measures* $\chi_S(G)$ and $\chi_S(J)$ *of chirality* are defined as

$$\chi_S(G) = 1 - \alpha_S(G) \tag{6.34}$$

and

$$\chi_S(J) = 1 - \alpha_S(J), \tag{6.35}$$

respectively. Objects with prominent chirality have χ_S measures close to 1, whereas achiral objects have χ_S measures equal to 0.

The above similarity based measures $\chi_S(G)$ and $\chi_S(J)$ of chirality are related to animals and polycubes of *maximal* chirality by the $\chi(J)$ and $\chi(G)$ criteria, since these are the very animals and polycubes having maximum chirality of 1 by the measures $\chi_S(G)$ and $\chi_S(J)$. However, the measures $\chi_S(G)$ and $\chi_S(J)$ of chirality are different from the measures $\chi(J)$ and $\chi(G)$. It is possible that for some curve J only chiral interior filling animals occur at some level n_C and above, yet the *same* chiral interior filling animal A of n_C cells and its mirror image $A^\Diamond$ may occur for both J and its mirror image $J^\Diamond$. Hence, at this level of resolution, the test of full dissimilarity fails, whereas the test of chirality by chiral interior filling animals already gives $n_\chi \leq n_C$. In such cases, the measures $\chi_S(G)$ and $\chi_S(J)$ provide more discrimination than $\chi(J)$ and $\chi(G)$.

Pictures of high resolution appear crisp, whereas pictures of low resolution appear fuzzy. A decrease of resolution is accompanied by an increase of fuzziness. Consequently, similarity measures based on the minimum level of resolution required to distinguish objects can be formulated in terms of the maximum level of fuzziness at which the objects are distinguishable. *Similarity can be regarded as fuzzy equivalence.* This principle provides an alternative mathematical basis for using the methods of *topological resolution* [262] in similarity analysis: the theory of fuzzy sets [382-385].

Fuzzy sets serve as mathematical tools for the description of problems where the classification criteria are not clearly defined. In ordinary set theory a point p is either an element or not an element of a set A, i.e., the *membership function* of a point p can have values of either 1 or 0. By contrast, in fuzzy set theory the membership function of a point p is regarded as the "degree of belonging" to the fuzzy set A, and this membership function can take any value from the interval [0,1]. In some sense, fuzzy set theory can be regarded as probability theory turned inside out, where events are considered with *"a posteriory"* certainty after they have occurred, but some uncertainty is associated with judging and classifying these events.

The theory of fuzzy sets [382-385] has numerous contemporary applications in various fields of engineering [386-388], in the description of quantum mechanical uncertainty [389-393], in the study of molecular identity preserving deformations [106,251], in new approaches to the description of approximate symmetry [252,394,395], as well as in both static and dynamic shape characterization and dynamic shape similarity analysis of molecules [55,396].

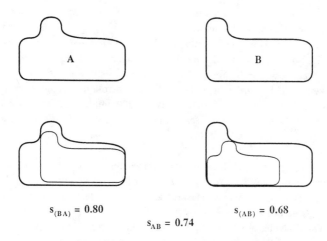

$$s_{(BA)} = 0.80 \qquad\qquad s_{(AB)} = 0.68$$
$$s_{AB} = 0.74$$

Figure   6.5   A two-dimensional example of the Scaling-Nesting Similarity Measure (SNSM),
applied to the planar objects A and B. A scaling of $s_{(BA)}$=0.80 is sufficient for fitting object B
within A,   whereas a scaling of $s_{(AB)}$=0.68 is needed for fitting object A within B. Note that,
according to these one-sided measures (semi-similarity measures), A is less similar to B than B is
to A.   A symmetric SNSM of s_{AB}=0.74 is obtained as the average of the above one-sided
scaling-nesting similarity measures.

In a general formulation, similarity measures taking values from the [0,1]
interval can be regarded as fuzzy set membership functions, expressing the degree of
belonging of one object to a class represented by the other object. On intuitive
basis one expects that such a relation is symmetric for any two objects: their degree
of belonging to the class of the other object is expected to be the same. However, this
need not be so, as illustrated by the asymmetric semi-similarity measures described
in the next section.

6.8   Semi-Similarity Measures and Scaling-Nesting Similarity Measures   (SNSM)

A general family of shape similarity measures of the second kind is based on the
principle of rescaling the size of one object until it fits within the other object. The
degree of scaling required gives a similarity measure.

These similarity measures are referred to as "Scaling-Nesting Similarity
Measures" (SNSM), and are illustrated in Figure 6.5. If the initial volumes of the
two objects are the same, then the scaling factor provides a numerical similarity
measure. In general, one expects that for objects of different shapes the larger the
scaling factor the greater their perceived similarity, and for objects of identical
shapes no scaling is required (i.e., the scaling factor is 1). Before the method can
be applied for objects of different initial volumes, one of the objects must be scaled

until its volume becomes identical to that of the other object. In the most general case, one object can be shifted as well as rotated with respect to the other while the fitting is attempted. However, various *constrained shape similarity measures* can be obtained if some motions are disallowed. For example, *orientation-dependent shape similarity measures* can be obtained if no rotations, only translations of the objects are allowed, or if no full rotation but only certain rotational angles are permitted while the fitting is attempted. Alternatively, if complete or limited rotation is allowed but the centers of the masses of the two objects are fixed or are restricted to limited domains of the space, then a *position-dependent similarity measure* is obtained.

The simplest cases of these similarity measures are not symmetric. In general, a different scaling may be required to fit A within B than that needed to fit B within A. That is, the above measures are *semi-similarity measures,* where the numerical value obtained depends on whether object A is tested against object B or *vice versa.* Note, however, that by taking the average of the two scaling factors, a symmetrized scaling-nesting similarity measure can be constructed.

For optimal fitting, rotation and translation of the objects are usually necessary. If an object A_v can be exactly superimposed on object A by translation and rotation (in general, by motions allowed by the actual orientation and position constraints), then we regard A_v as a *version* of object A.

A definition of the unconstrained scaling-nesting similarity measures is given below; the constrained measures can be derived easily from these by applying the appropriate constraint while determining the maximum scaling factor.

Consider two bounded and closed 3D objects, A and B°. If V(A) and V(B°) denote the volumes of these objects, then first a uniform scaling

$$B = s(B°) \tag{6.36}$$

is applied to B° so that the volume of the new object B is equal to that of A:

$$V(A) = V(B). \tag{6.37}$$

The uniform scaling s(B°) can be characterized by a single scaling factor s:

$$V(B) = s^n V(B°). \tag{6.38}$$

In the above equations, the same letter symbol s is used to denote the scaling operation (in Equation (6.36)), and the numerical value of the scaling factor (in Equation (6.38)). In general, the n-dimensional volumes are scaled by the n-th power s^n of the scaling factor s. In the example of Figure 6.5, the dimension is n=2, and the 2D "volumes" (areas) are scaled by the squares of the scaling factors.

Note that the shapes of B° and B are regarded identical; these two objects differ only in size.

In the next step, additional scalings $s_A(A)$ and $s_B(B)$ are carried out on all possible rotated and translated versions A_v and B_v of objects A and B, leading to two scalar numbers $s_{(AB)}$ and $s_{(BA)}$:

$$s_{(AB)} = \max \{\; s_A : \text{exists } A_v \text{ such that } s_A(A_v) \subset B \}, \tag{6.39}$$

$$s_{(BA)} = \max \{\; s_B : \text{exists } B_v \text{ such that } s_B(B_v) \subset A \}. \tag{6.40}$$

The above two numbers $s_{(AB)}$ and $s_{(BA)}$ are *scaling-nesting semi-similarity measures*, or *one-sided scaling-nesting similarity measures*, $s_{(AB)}$ expressing the semi-similarity of A to B and $s_{(BA)}$ expressing the semi-similarity of B to A.
   A *symmetric scaling-nesting similarity measure* is obtained by

$$s_{AB} = (s_{(BA)} + s_{(BA)}) / 2 \;. \tag{6.41}$$

Alternatively, one may take the greater (or the smaller) of the one-sided similarity measures for any two objects; this method also provides a symmetrized version of the one-sided similarity measures.
   These similarity measures of the second kind are applicable to mirror images. The quantity $s_{AA^\lozenge}$ can be taken as a measure of achirality, leading to a new measure of chirality:

$$\chi_{SAB}(A) = 1 - s_{AA^\lozenge} \;. \tag{6.42}$$

The quantity $\chi_{SAB}(A)$ is zero for achiral objects, and it is a positive value for chiral objects. In general, $\chi_{SAB}(A)$ tends to have larger values for objects perceived as having more prominent chirality. This measure differs from measures based on maximum overlap between mirror images [46,48-53,57,58,242].

6.9   Molecular Similarity Measures Based on Shape Codes

The comparison of numerical shape codes based on the curvature domain partitionings of molecular surfaces provides a natural definition for similarity measures. One of the simplest of these is based on the shape groups, where the (a,b) parameter maps generated by the shape groups are compared. The shape groups distributed along an (a,b) parameter map can be characterized by numbers, for example, by their Betti numbers, and the entire map can be represented by a sequence of numbers ordered into a matrix or a vector. The same method applies if each separate piece of the molecular surface obtained after truncation is regarded as a separate entity with its separate group [192,262,263]. In this case, the Betti number of each piece is specified separately, and the set of these Betti numbers is assigned to the given location of the (a,b) parameter map.
   A simple coding technique has been based on an ordering of the separate pieces of the truncated molecular surface according to their decreasing size ([193], see also [109]). For a given molecule, different MIDCO surfaces G(a) are obtained for different density threshold parameters a, and a different curvature criterion b leads to a different curvature domain partitioning of the MIDCO. Consequently, the truncation of MIDCO's of various threshold density values using various curvature criteria may lead to a great variety in both the number and the actual

shape properties of the pieces, hence in most cases, the size ordering of the pieces must be repeated for each selected (a,b) parameter pair. For each given (a,b) pair, the size-ordering of the surface pieces implies an ordering of their one-dimensional Betti numbers (informally called their "first" Betti numbers) into a sequence:

$$B(1), B(2), \ldots, B(k), \ldots, B(m). \tag{6.43}$$

Usually, the surface piece with the largest Betti number is the one with the largest surface area, implying that the sequence of Betti numbers in the above ordering is approximately the same as the *decreasing sequence* of the Betti numbers.

Such a decreasing sequence is associated with each point of a grid on the (a,b) parameter map. The grid itself can be regarded as a matrix with the sequences of Betti numbers as elements. This matrix is a *shape code* for the molecule that can be used for storing shape information in molecular data banks and can be retrieved for shape similarity assessment by numerical methods.

In the following paragraphs alternative numerical representations of these shape codes are described, where the shape information is given by a "shape ID vector" of a specified dimension, or it is compressed into a single number that can be regarded as a "shape ID number" of the given molecular conformation. For larger molecules these shape ID numbers often turn out very large and somewhat clumsy for practical applications, however, they are of some theoretical importance.

In general, for different (a,b) pairs, both the number m of Betti numbers, and the actual value of the Betti numbers may change. Consequently, coding methods relying on simple listings of the sequences (6.43) for each selected (a,b) pair lack uniformity, since lists of greatly varying lengths are to be coded. Although such direct coding methods are simple and useful, it is advantageous to use methods which are uniform for all (a,b) pairs.

For each parameter pair (a,b), a single number $c'(a,b)$ can be used to store the information on the entire sequence (6.43) of ordered Betti numbers. One coding-decoding method relies on the prime factorization of integers. If p_i denotes the i-th prime number in the sequence

$$1, 2, 3, 5, 7, 11, 13, 17, \ldots \ldots \tag{6.44}$$

of primes, then the code $c'(a,b)$ is defined as the following product:

$$c'(a,b) = 2^{B(1)+1} \times 3^{B(2)+1} \times \cdots \times (p_{k+1})^{B(k)+1} \times \cdots \times (p_{m+1})^{B(m)+1}. \tag{6.45}$$

This single number $c'(a,b)$ contains all information present in the values and the ordering of the original sequence of Betti numbers. The code $c'(a,b)$ can be decoded easily by virtue of the prime factorization theorem: the code $c'(a,b)$ has a unique representation as a product of primes. For the given value $c'(a,b)$, the exponents $r(k+1)$ obtained in the prime factorization

$$c'(a,b) = 2^{r(2)} \times 3^{r(3)} \times \cdots \times (p_{k+1})^{r(k+1)} \times \cdots \times (p_{m+1})^{r(m+1)} \tag{6.46}$$

of the code c'(a,b) provide the set (6.43) of original Betti numbers which can be computed by the following simple relation:

$$B(k) = r(k+1) - 1.$$ (6.47)

For example, consider a given electron density threshold and reference curvature parameter pair (a,b), and assume that the associated truncated contour surface falls into seven pieces. Furthermore, assume that a shape group analysis gives the sequence

$$6, 3, 3, 0, 0, 0, 0$$ (6.48)

for the corresponding Betti numbers, ordered according to the decreasing size of the areas of the surface pieces. Note that a decreasing sequence of surface areas is not necessarily accompanied by a decreasing sequence of Betti numbers, although such parallel trends are common, as indicated by the example. For this example, the code is

$$\begin{aligned} c'(a,b) &= 2^{6+1} \times 3^{3+1} \times 5^{3+1} \times 7^{0+1} \times 11^{0+1} \times 13^{0+1} \times 17^{0+1} \\ &= 110270160000. \end{aligned}$$ (6.49)

This numerical code, representing shape information for the given (a,b) parameter pair, is assigned to the associated (a,b) point of the (a,b) parameter map.

When needed, this single integer c'(a,b) can be decoded. Of course, the number 110270160000 has a unique prime factorization,

$$110270160000 = 2^7 \times 3^4 \times 5^4 \times 7 \times 11 \times 13 \times 17,$$ (6.50)

and using the relation $B(k) = r(k+1) - 1$, and the actual values of the exponents $r(k+1)$, the Betti numbers, as well as their ordering, are easily calculated. For example, 7 is the fifth prime number,

$$7 = p_5,$$ (6.51)

and the prime factor of 7 occurs on the first power in the number 110270160000, consequently,

$$r(5) = 1.$$ (6.52)

The above result implies that the surface piece of serial number 5-1 = 4 according to decreasing size has a Betti number equal to 1-1 = 0,

$$B(4) = r(5) - 1 = 1 - 1 = 0.$$ (6.53)

This method encodes some size information in addition to shape. Note, however, that the method often leads to large numbers for the code c'(a,b). For large numbers, the prime factorization can become somewhat time consuming; hence, the decoding process is somewhat less efficient than in alternative methods which ignore size information.

One such alternative coding method is described below, where the Betti numbers are ordered into a sequence

$$B(1), B(2), \ldots, B(j), \ldots, B(m). \hspace{3cm} (6.54)$$

according to their *increasing* magnitude [263], without consideration of the size of the surface pieces they represent. A single number $c(a,b)$ can be used to store this information. If p_i denotes the i-th prime number as before, then the code $c(a,b)$ is defined as the following product:

$$c(a,b) = p_{B(1)+2} \times p_{B(2)+2} \times \cdots \times p_{B(j)+2} \times \cdots \times p_{B(m)+2}. \hspace{1cm} (6.55)$$

If a Betti number $B(j)$ occurs t times in the sequence, then it is represented by a factor $(p_{B(j)+2})^t$, and the code $c(a,b)$ can be taken as the product of all these factors.

The number $c(a,b)$ can be decoded easily. By virtue of the prime factorization theorem, the original set of Betti numbers can be calculated from the prime factors of the number $c(a,b)$. If the prime factorization gives

$$c(a,b) = (p_{i(1)})^{t(1)} \times (p_{i(2)})^{t(2)} \times \cdots \times (p_{i(s)})^{t(s)} \times \cdots \times (p_{i(w)})^{t(w)}, \hspace{0.5cm} (6.56)$$

then the Betti numbers which occur are

$$B(j) = i(s) - 2, \hspace{5cm} (6.57)$$

where

$$t(1) + t(2) + \ldots + t(s-1) < j \le t(1) + t(2) + \ldots + t(s-1) + t(s). \hspace{0.5cm} (6.58)$$

For the example considered above, the sequence is

$$0, 0, 0, 0, 3, 3, 6. \hspace{5cm} (6.59)$$

For this example, the evaluation of the code $c(a,b)$ according to the definition (6.55) gives

$$c(a,b) = 2^4 \times 7^2 \times 17 = 13328. \hspace{3cm} (6.60)$$

This value is taken as the code for the set $0, 0, 0, 0, 3, 3, 6$ of Betti numbers. The sequence $0, 0, 0, 0, 3, 3, 6$ can be recovered easily from the unique prime factors of the number 13328. One should note that this number is much smaller than the number 110270160000 of the $c'(a,b)$ code described above; hence, in general, the prime factorization in the decoding step takes much less time. However, this code does not contain direct size information, although the approximate parallel trend in the magnitudes of the Betti numbers and surface areas can be used as a less than fully reliable guideline.

Both of the above coding techniques compress several integers into a single integer number that can be decoded by prime factorization. The advantage of these techniques lies in the fact that the same method is applicable to all choices of a and b values, allowing large variations in the length of the sequence of the Betti numbers.

For either code, $c(a,b)$ or $c'(a,b)$, the corresponding number can be assigned to the (a,b) location of the parameter map (a,b). Since most small changes in the values of a and b do not change the shape groups of the actual truncated molecular surfaces, the entire (a,b) map will contain only a finite number of different values for the $c(a,b)$ or $c'(a,b)$ code. A list of these code values can be regarded as a vector, providing a numerical shape code for the entire (a,b) map (i.e., for all relevant electron density values a and test curvature values b).

For practical purposes, it is useful to consider a rectangular grid on the (a,b) map. In one implementation [263], a 41×21 grid is considered, covering a range of [0.001 - 0.1 a.u.] (a.u. = atomic unit) of density threshold values a, and a curvature range of [(-1),1.0] for the test spheres against which the local curvatures of the MIDCO are compared. At each of these grid points the $c(a,b)$ code or the $c'(a,b)$ code is calculated, and the actual shape code of the entire 3D electron density is taken as the resulting 41×21 matrix $\mathbb{C}$ of integers, stored either as a matrix or as an integer vector $\mathbf{C}$ of 861 components.

A simple numerical similarity measure is defined by the number of matches between the components of two such shape code vectors. If $\mathbf{C}(M_1)$ and $\mathbf{C}(M_2)$ are the shape code vectors of molecules (or conformers) M_1 and M_2, then the similarity measure $s_C(M_1, M_2)$ is defined as

$$s_C(M_1, M_2) = \sum_{i=1}^{861} \delta_{j(i),k(i)} / 861, \qquad (6.61)$$

where $\delta_{j,k}$ is the Kroenecker delta, with indices

$$j(i) = C_i(M_1), \text{ and } k(i) = C_i(M_2), \qquad (6.62)$$

that is, $s_C(M_1, M_2)$ is the number of matches divided by the dimension 861 of the code vectors $\mathbf{C}(M_1)$ and $\mathbf{C}(M_2)$.

A somewhat cruder shape similarity measure is obtained by considering the sequence of topologically different sets of density domains (DD) occurring as the electronic density threshold value a is gradually decreased [262]. For two molecules (or conformers) M_1 and M_2, the maximum number of matches along their sequences, divided by the number of different sets in the longer sequence provides a numerical similarity measure [262]. Since most topological changes of density domains occur at high electron density thresholds, this similarity measure focuses on the relatively high values of electron density, whereas the more versatile measure based on the shape codes $c(a,b)$ or $c'(a,b)$ discussed above allows an optional choice for the density interval considered.

One of the most useful shape codes is based on shape matrices. As we have seen in Chapter 5, the N-neighbor relation $N(D_{\mu,i}, D_{\mu',i'})$ of various curvature domains $D_{\mu,i}$ and $D_{\mu',i'}$, given by Equation (5.8), leads to a shape matrix representation of the shape of the molecular surface. The shape matrix $s(a,b)$ depends on the two parameters a and b. The off-diagonal elements of this matrix $s(a,b)$ are defined as $s(a,b)_{i,i'} = N(D_{\mu,i}, D_{\mu',i'})$ and its diagonal elements $s(a,b)_{i,i} = \mu_i(a,b)$ are the $\mu(r,b)$ indices of points r within the ith shape domain $D_{\mu,i}$ of the MIDCO $G(a)$.

The assignment of the index i to the curvature domains $D_{\mu,i}$ can be used to encode additional information; for example, some indication of relative sizes of the shape domains. In one implementation [109], the index i follows the ordering of all the $D_{\mu,i}$ shape domains according to the decreasing size of their surface areas on the MIDCO $G(a)$. In this case, the shape matrix $s(a,b)$ encodes both shape and size information. The comparison of molecular shapes (and to some extent, molecular sizes) can be accomplished by a comparison of their shape matrices.

The corresponding shape code vector $c(s(a,b))$, a three-dimensional vector, is constructed in three steps.

1. The first component $c_1(s(a,b))$ of the shape code vector $c(s(a,b))$ is the dimension n of the $n \times n$ matrix $s(a,b)$.

2. The n diagonal elements $\mu_i(a,b)$ of the matrix $s(a,b)$ are encoded by concatenating them into a single number. For example, a diagonal of elements $(2, 2, 1, 2, 0, 2)$ is encoded as the decimal number 221202. (Since the only numerical values that occur along the diagonal are 0, 1, and 2, a ternary number system can also be used). The number so obtained is the second component $c_2(s(a,b))$ of the shape code vector $c(s(a,b))$.

3. The upper off-diagonal triangle of the shape matrix has elements 1 and 0 only, and these $n(n-1)/2$ numbers concatenated according to columns form a binary number that is the third component $c_3(s(a,b))$ of the shape code vector $c(s(a,b))$.

The resulting code vector $c(s(a,b))$ is assigned to the (a,b) point of the parameter map (a,b).

Note that if the element $s(a,b)_{1,1}$ of the shape matrix is different from zero, then the information on the dimension n of the matrix can be deduced from the second element $c_2(s(a,b))$ of the shape code vector $c(s(a,b))$. The special case of $s(a,b)_{1,1} = 0$ seldom occurs, since this implies that the largest D_μ domain is a locally concave D_0 domain relative to the curvature parameter b. Nevertheless, in order to avoid ambiguity in such cases, the dimension n is specified as the first component $c_1(s(a,b))$ of the shape code vector $c(s(a,b))$.

Also note that concatenation of the upper off-diagonal triangle of the shape matrix by columns instead of rows ensures that shape matrices of different dimensions can be easily compared: if a k-th digit is present in both of the resulting binary numbers, then they correspond to the same pair of row and column indices in the two shape matrices. Evidently, this feature is of importance when comparing two molecules, and also when comparing the shapes of two different conformers of the same molecule: in both cases the dimensions of the shape matrices may differ.

As an example, the shape matrix $s(0.01,0)$ of the b=0 shape domain partitionings of the allyl alcohol MIDCO $G(0.01)$ shown in Figure 5.6 belongs to the $(0.01,0)$ point of the parameter map (a,b). If the index ordering of the various $D_{\mu,i}$ domains follows the order of decreasing size of their surface area, then the shape matrix $s(0.01,0)$ has the form given by Equation (5.12):

$$s(0.01,0) = \begin{matrix} 2 & 1 & 0 & 1 & 0 & 0 & 1 & 0 \\ 1 & 1 & 1 & 0 & 1 & 1 & 0 & 0 \\ 0 & 1 & 2 & 0 & 0 & 0 & 0 & 1 \\ 1 & 0 & 0 & 1 & 0 & 0 & 0 & 0 \\ 0 & 1 & 0 & 0 & 2 & 0 & 0 & 0 \\ 0 & 1 & 0 & 0 & 0 & 2 & 0 & 0 \\ 1 & 0 & 0 & 0 & 0 & 0 & 1 & 0 \\ 0 & 0 & 1 & 0 & 0 & 0 & 0 & 1 \end{matrix}$$

The first element of the corresponding shape vector $c(s(a,b))$ is the dimension 8 of the shape matrix:

$$c_1(s(0.01,0)) = 8. \tag{6.63}$$

The digits of the number $c_2(s(a,b))$ are the $\mu_i(a,b)$ diagonal elements of the shape matrix $s(0.01,0)$:

$$c_2(s(0.01,0)) = 21212211. \tag{6.64}$$

In order to determine the third component $c_3(s(0.01,0))$ of the code vector $c(s(a,b))$, the elements of the upper off-diagonal triangle of the shape matrix $s(0.01,0)$ are concatenated column-continuously. This results in the binary number 10110001000100010000000010000, equal to the ten-base number of 371335200, hence

$$c_3(s(0.01,0)) = 371335200. \tag{6.65}$$

That is, the code vector $c(s(a,b))$ for the b=0 curvature analysis of the allyl alcohol MIDCO $G(0.01)$ is

$$c(s(0.01,0)) = (8, 21212211, 371335200). \tag{6.66}$$

This shape code vector can be decoded easily by simply reversing the above process. In view of the relation $n = c_1(s(0.01,0))$, the reconstruction of the diagonal elements of matrix $s(0.01,0)$ from $c_2(s(0.01,0))$ is a trivial task, whereas the conversion of the third component $c_3(s(0.01,0))$ into a binary number of digits d_r leads to the off-diagonal elements $s(0.01,0)_{i,j} = d_r$ of shape matrix $s(0.01,0)$, where the relations $r = i+(j-1)(j-2)/2$ and $i < j$ assign a unique pair of indices i,j to each index r.

In practice, only a finite number of parameter pairs (a,b) are considered, for example, those at the grid points of the 41×21 grid of the parameter map (a,b) described above. The entire map of shape matrix codes can then be represented by three 861-dimensional vectors, $\mathbf{C}^{(1)}$, $\mathbf{C}^{(2)}$, and $\mathbf{C}^{(3)}$, containing all first, all second, and all third components, respectively, of the individual $\mathbf{c}(s(a,b))$ vectors. Alternatively, a single (3×861) - dimensional vector $\mathbf{C}$ can be assigned to the (a,b) parameter map, where $\mathbf{C}$ is obtained by concatenating the components of $\mathbf{C}^{(1)}$, $\mathbf{C}^{(2)}$, and $\mathbf{C}^{(3)}$ into a single vector.

A similarity assessment of the shapes encoded by these vectors can be carried out on various levels. For two molecules (or conformers) M_1 and M_2, the number of matches along their $\mathbf{C}(M_1)$ and $\mathbf{C}(M_2)$ vectors, divided by the dimension (3×861), provides a simple, numerical similarity measure [262]. The comparison of the first 2×861 components [i.e., the comparison of vectors $\mathbf{C}^{(1)}(M_1)$ and $\mathbf{C}^{(2)}(M_1)$ to $\mathbf{C}^{(1)}(M_2)$ and $\mathbf{C}^{(2)}(M_2)$, respectively] appears the most important for a crude shape analysis, since these vectors store the information on the number and type of shape domains occurring for various a and b parameter values. By contrast, a direct comparison of vectors $\mathbf{C}^{(3)}(M_1)$ and $\mathbf{C}^{(3)}(M_2)$ gives information on the similarities of the patterns of different shape domains on the two families of molecular surfaces.

More detailed shape comparison is possible if the *decoded* elements of the two vectors $\mathbf{C}(M_1)$ and $\mathbf{C}(M_2)$ are compared directly. For example, by taking the number of matches along the diagonals and within the off-diagonal upper triangles of the two shape matrices $s(a,b,M_1)$ and $s(a,b,M_2)$, divided by $n(n+1)/2$, where n is the dimension of the larger of the two matrices, an elementary similarity measure s(a,b) is obtained, characteristic to the point (a,b) of the parameter map. Clearly,

$$0 \leq s(a,b) \leq 1. \tag{6.67}$$

These s(a,b) values generate a similarity map over the (a,b) parameter plane. These maps, or on a simpler level, the average, minimum, and maximum s(a,b) values provide more general similarity measures,

$$s_a(M_1,M_2), \tag{6.68}$$

$$s_m(M_1,M_2), \tag{6.69}$$

and

$$s_M(M_1,M_2), \tag{6.70}$$

respectively.

In the above discussions, a size ordering of the shape domains has been considered. It is possible that a better overall match is achieved if deviations from a strict size ordering are allowed. In the most general case, all simultaneous row and column permutations of the larger shape matrix are considered, and the actual permutation providing the best match is selected.

A rather general similarity measure is provided by a comparison of the vertex labeled graphs of shape globe invariance maps (SGIM's). As discussed in Section 5.4, these maps are applicable to molecular contour surfaces, e.g., MIDCO's or MEPCO's, and also for chain molecule backbones, such as those used in the study of the folding patterns of proteins. The number of elementary changes required to turn one such graph into another one gives an indication of their dissimilarity. The elementary changes, that is, the elementary vertex labeled graph operations are the following: addition or deletion of a vertex, addition or deletion of an edge, and a change of vertex label. Accordingly, a similarity measure for two such SGIM's is defined as

$$s_{SGIM}(M_1, M_2) = 1 - c/\max(2v+e), \tag{6.71}$$

where c is the minimum number of elementary vertex labeled graph operations needed to convert the vertex labeled graph of one map to that of the other map, and v and e are the number of vertices and the number of edges, respectively, of the graph considered. Note that the larger of the two possible sums $(2v+e)$ is used in the definition.

6.10   Local Shape Codes and Local Similarity Measures

In many chemical problems the comparisons of local molecular regions are more important than global comparisons. The presence of functional groups or other molecular moieties with specified shape properties often imply similar chemical behavior even if the molecules compared have very different global shapes. For this reason, local molecular shape descriptors and local shape codes are of major importance.

The general methods applied for shape codes based on shape matrices are especially suitable for developing local shape codes.

For a given choice of parameters a and b, a local moiety of molecule M_1 involves only a subfamily of the family of $D_{\mu,i}$ shape domains of the molecular surface. For example, if the moiety corresponds to a DD functional group, then at some density threshold a separate DD is assigned to this moiety. This implies that even in lower-density ranges where this DD is already joined to other DD's, the curvature domains which belong primarily to the moiety are distinguishable from other curvature domains of the MIDCO's. Let us assume that for the given (a,b) parameter pair there are k shape domains in the subfamily corresponding to the molecular moiety. These k domains may have very different sizes and the indices i of size ordering of members of this subfamily do not necessarily appear consecutively along the diagonal of the shape matrix $s(a,b,M_1)$. In such cases, simultaneous row and column permutations in the shape matrix $s(a,b,M_1)$ may be needed in order to arrange these shape domains consecutively while preserving the size ordering among themselves. We assume that the actual ordering in the shape matrix $s(a,b,M_1)$ fulfills these criteria. Then, the shape of the local molecular moiety can be represented by a k-dimensional diagonal block $b(a,b,M_1)$ of the n-dimensional global shape matrix $s(a,b,M_1)$ of molecule M_1.

The very same procedure that has been used to construct the global $C(M_1)$ codes for the global shape matrices $s(a,b,M_1)$ along the parameter map (a,b) can also be applied to the set of local shape matrices $b(a,b,M_1)$ along the parameter map (a,b), resulting in a local shape code vector

$$C_L(M_1) \tag{6.72}$$

of the given molecular moiety.

When searching for local similarities of two molecules, the decoded local shape matrices $b(a,b,M_1)$ of molecule M_1 are compared to various diagonal blocks of the global shape matrix $s(a,b,M_2)$ of molecule M_2. In the most general case, the local shape matrix $b(a,b,M_1)$ is used as a template, and it is compared to k-dimensional blocks of $s(a,b,M_2)$ obtained by all possible simultaneous row and column permutations. If the size ordering is considered important then only those permutations are taken which preserve the monotonicity of size ordering in the permuted diagonal block that is compared to the template. A local similarity measure

$$s_L(M_1,M_2) \tag{6.73}$$

is the ratio of the number of matches to the total number of entries in the local shape codes.

The global and local shape codes can be used for measuring *global and local shape compexity,* respectively. Let $w(s(a,b,M))$ and $w(b(a,b,M))$ denote the number of different entries of the n-dimensional global shape matrix $s(a,b,M)$ and a k-dimensional local shape matrix $b(a,b,M)$, respectively. Simple global and local shape complexity measures of molecule M are defined as the following ratios:

$$x_{global}(M) = w(s(a,b,M)) / n , \tag{6.74}$$

and for the local molecular moiety specified by the k-dimensional block $b(a,b,M)$,

$$x_{local}(M) = w(b(a,b,M)) / k . \tag{6.75}$$

6.11   Molecular Shape Complementarity Measures

In nearly all chemically important problems, shape complementarity refers to local shape properties. Most of the typical molecular interactions where shape complementarity is relevant involve only some local moieties of the molecules. Global shape complementarity is more difficult to achieve and seldom plays a role,

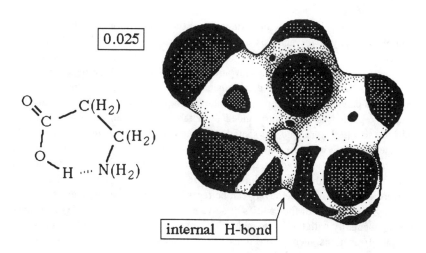

Figure   6.6   A toroidal MIDCO G(a) of a = 0.025 a.u. (atomic unit) of one of the stable conformers of the β-alanine molecule. The torus is completed by a hydrogen bond formed between the terminal N atom and the OH group. The $D_{\mu,i}$ shape domains indicated are those corresponding to the b=0 reference curvature parameter (ordinary convexity). The torus is much narrower along the hydrogen bond than along formal single bonds of the ring. The hole of the torus is rather small, hence the topology of any surrounding chain molecule is unlikely to complement the correct toroidal topology of the enclosed conformer of the β-alanine molecule.

one exception is the special case of one molecule, such as a long, folded chain molecule entirely surrounding another, usually simpler molecule.

A properly folded chain molecule can form a cavity with a shape globally complementing the shape of an enclosed, quasi-spherical molecule. If, however, the enclosed molecule has a toroidal or more complicated topology, where the hole of the torus is small so the surrounding chain molecule cannot enter this hole, then an approximate global complementarity by a chain molecule is unlikely to reflect the correct topology of the enclosed molecule.

Typically, global shape complementarity for molecules with toroidal MIDCO's can be achieved only to a much lesser degree than for quasi-spherical molecules. An example of a toroidal MIDCO where the torus is completed by an internal hydrogen bond is shown in Figure 6.6. In this example, the hydrogen bond is formed between the terminal N atom and the OH group of one of the stable conformers of the β-alanine molecule [263]. The $D_{\mu,i}$ shape domains indicated are those corresponding to the b=0 value of the reference curvature parameter (case of ordinary convexity). Although the torus is much narrower between the H and N nuclei of the hydrogen bond than along formal single bonds of the ring, nevertheless, the hydrogen bond is clearly recognizable. The hole of the torus is rather small, hence approximate global shape complementarity by any surrounding chain molecule is unlikely to reflect the correct toroidal topology of the enclosed β-alanine molecule.

There are, however, important potential applications of global shape complementarity if the analysis is not restricted to a molecule pair. Clearly, one molecule can be surrounded by several other molecules and their mutual arrangements are expected to be favourable in the case of a high degree of shape complementarity. Such shape complementarity of one molecule with a family of molecules plays an important role in solute-solvent interactions. A solvated molecule can be regarded as being surrounded by a family of solvent molecules forming various solvate layers, and the spatial arrangements of solvent molecules in these solvate layers are influenced by the shape of the solute molecule. Solvate layers are dynamic entities undergoing continuous rearrangements, furthermore, hollow solvate layers without the solute molecule seldom have any stability at all. Nevertheless, these solvate layers can be regarded as molecular aggregates of some identity. The shapes of the cavities of solvate layers and the degree of global shape complementarity with the solute molecules provide important insight. In liquid phase chemical reactions such as reactions in the aqueous phase, the reactants usually undergo major shape changes until they reach the shapes of the product molecules. These shape changes of the reacting molecules are approximately mirrored in the complementary shape changes of the solvate layers. However, this complementarity is not uniform in all stages of the reaction; the degree of complementarity may change during the process. It is expected that the mediating effect of solvents in liquid phase reactions is strongly influenced by the changes of the degree of shape complementarity of solvate layers throughout the reaction.

The catalytic activity of zeolites also involves shape complementarity. The cavities of zeolites are interconnected by various channels, consequently, in a strict sense, a cavity does not fully surround a molecule that enters the zeolite. As a result, for high density zeolite MIDCO's only local shape complementarity is possible. Nevertheless, the low density MIDCO's of zeolites contain closed internal contour surfaces corresponding to the cavities and for these parts of MIDCO's global shape complementarity is relevant. The shapes of these MIDCO's may approximately complement the shapes of the MIDCO's of a molecule inside the cavity.

For the simpler case of molecule pair interactions, such as the interactions between two reactants, local shape complementarity is of importance. The basic principle of local shape similarity measures is also applicable for the construction of local shape complementarity measures.

Shape complementarity of molecular electron densities represented by MIDCO's involves complementary curvatures, as well as complementary values of the charge density contour parameters a. In general, a locally convex domain relative to a reference curvature b shows shape complementarity with a locally concave domain relative to a reference curvature -b. Furthermore, shape complementarity between the lower electron density contours of one molecule and the higher electron density contours of the other molecule is of importance.

In the course of molecular interactions, the interacting molecules penetrate each other only to a limited extent. For stronger interactions, this interpenetration is assumed to be greater than that for weaker interactions. For the given interaction, one may consider a *common* electronic density value a_0 of the two isolated molecules that corresponds to the threshold density of two formal "contact" MIDCO's of the interacting systems. If, in a crude model, the electronic density of the interacting pair of molecules M_1 and M_2 is approximated by the superposition of the electronic densities of the two isolated molecules, then a pair of MIDCO surfaces $G(a, M_1)$ and $G(a, M_2)$ of the same a value can have one of the three possible relative arrangements:

1. they have no common points, or
2. they have a finite number of common points (usually, one common point), or
3. they have a continuum of common points.

For the given mutual arrangement of the two molecules M_1 and M_2, the *contact density* a_0 corresponds to the unique electron density threshold value of the MIDCO's of case 2.

If in an approximate model of molecular interactions a contact density value a_0 can be chosen, then the local shape complementarity between $G(a_0, M_1)$ and $G(a_0, M_2)$ is of relevance. In a more general model, one considers the local shape complementarity of MIDCO's $G(a_0-a', M_1)$ and $G(a_0+a', M_2)$ in a narrow density interval

$$[a_0 - \Delta a, \ a_0 + \Delta a]. \tag{6.76}$$

In this model, the complementarity of the local shapes of MIDCO's deviating in the *opposite sense* from the contact density value a_0 is analysed.

Complementarity involves matches between locally concave and locally convex domains, as well as matches between saddle-type domains with proper alignment. Taking into account the conditions for density thresholds and curvature parameters, the task is to find local matches between curvature domain pairs

$$D_{0(b),i}(a_0 - a', M_1), \ D_{2(-b),i}(a_0+a', M_2); \tag{6.77}$$

$$D_{1(b),i}(a_0 - a', M_1), \ D_{1(-b),i}(a_0+a', M_2); \tag{6.78}$$

and

$$D_{2(b),i}(a_0 - a', M_1), \ D_{1(-b),i}(a_0+a', M_2). \tag{6.79}$$

The above considerations can be incorporated within a simple model using the (a,b) parameter map approach. The complementarity of curvature domain types (for example, D_2 and D_0) can be tested by taking complementary shape groups, that is, by taking complementary truncations for molecule M_1 and M_2. In general, this leads to an (a,b) map for the $HP_\mu(a,b)$ shape groups of molecule M_1 and to an (a,b) map for the complementary $HP_{2-\mu}(a,b)$ shape groups of molecule M_2. For example, one may take the most useful one-dimensional shape groups $(p = 1)$ for both molecules, with reference to the $\mu = 2$ truncation for molecule M_1, and to the complementary $\mu' = 2-\mu = 0$ truncation for molecule M_2. This choice leads to the (a,b) map of the $H^1_2(a,b)$ shape groups of molecule M_1 and to the (a,b) map of the $H^1_0(a,b)$ shape groups of molecule M_2.

Whereas the curvature types for truncation are complementary, the above two (a,b) maps cannot yet be compared directly, since in a direct comparison of these maps, identical, and not complementary, a and b values occur for the two molecules. However, the complementarity of density thresholds and curvatures can be taken into account by a simple transformation: *by inverting the* (a,b) *parameter map of molecule* M_2 *centrally with respect to the point* $(a_0,0)$, and by comparing the centrally inverted (a,b) map of M_2 to the original (a,b) map of M_1. This transformation ensures that domain types, density thresholds, and curvature parameters are matched properly, as required by the pairing scheme $(6.77) - (6.79)$. For example, the locally convex domains of MIDCO $G(a_0-a', M_1)$ relative to the reference curvature b are tested for shape complementarity against the locally concave domains of MIDCO $G(a_0+a', M_2)$ relative to a reference curvature $- b$.

This Centrally Inverted Map Method (CIMM) of molecular shape complementarity analysis allows one to use the techniques of similarity measures. In fact, *the problem of shape complementarity is converted into a problem of similarity* between the original (a,b) parameter map of shape groups $HP_\mu(a,b)$ of molecule M_1 and the centrally inverted (a,b) parameter map of the complementary $HP_{2-\mu}(a,b)$ shape groups of molecule M_2.

For most practical applications, CIMM is used within the framework of local measures. These measures are based on local shape matrices or on the shape groups of local moieties, defined either by the density domain approach mentioned earlier, or by alternative conditions, such as the simple truncation condition replacing the "remainder" of the molecule by a generic domain [192]. For proper complementarity, identity or close similarity of the patterns of the matched domains is an advantage, hence the parts $C_L^{(3)}(M_1)$ and $C_L^{(3)}(M_2)$ of the corresponding local shape codes are compared directly. For shape complementarity only the specified density range $[a_0 - \Delta a,\ a_0 + \Delta a]$ and a specified curvature range of the (a,b) parameter maps is considered. A local shape complementarity measure, denoted by

$$c_L(M_1,M_2), \hspace{4cm} (6.80)$$

is analogous to the local version $s_L(M_1,M_2)$ of the similarity measure given by Equation (6.61). This local shape complementarity measure $c_L(M_1,M_2)$ is defined as the number of matches between the entries of maps obtained by CIMM, divided by the total number of comparisons.

CHAPTER
7

QShAR
(QUANTITATIVE SHAPE - ACTIVITY RELATIONS)
IN DRUG DESIGN AND MOLECULAR ENGINEERING

In the continuing effort to understand the reactions of biomolecules and to design new drugs for medicine, large data bases of most major pharmaceutical drug companies are routinely searched for structural motifs as defined by functional groups, bonding patterns, and substructures of molecular structural diagrams. Such data bases provide input information for molecular modeling. In particular, the *integration* of experimental data with computer-based molecular modeling has become a powerful tool for the interpretation of biochemical processes, drug design, and molecular engineering (see, e.g., references [85-88,160-198,282-341]). A systematic survey and handling of the available structural data of molecular bonding patterns, such as that available in the discrete representations of the DARC system [397-406], allows one easy access to such information. These techniques also pinpoint the areas where our knowledge is lacking, which is an important consideration when establishing research goals.

Molecules of similar biochemical activity often show common 3D shape features. Consequently, the characterization of the shapes of formal molecular bodies and the recognition, description, and, ultimately, the numerical evaluation of similarity among molecules are of major importance in modern pharmaceutical research, as well as in pesticide and herbicide chemistry. The analysis of molecular shape is an important component of research aimed at the elucidation of drug-receptor interactions and in studies of quantitative structure-activity relationships in contemporary drug design.

The molecular data bases organized according to *chemical formulas* and *bonding patterns* are complemented by molecular data bases which can be searched using the *three-dimensional shapes of formal molecular bodies* as criterion. Molecular shape effects, with particular emphasis on the 3D *body* aspect of molecules, as opposed to the *skeletal* aspect of steric arrangements of formal bonds, are of primary importance in biochemical processes (e.g., in enzyme-drug interactions). Shape similarity and shape complementarity are fundamental principles of biochemistry, relevant to the properties of both stable macromolecular structures and transition structures of biochemical reactions. In particular, the suggestion of Linus Pauling concerning the shape complementarity of the enzyme active site and the transition structure of the reaction catalyzed by the enzyme gives important insight.

In most chemical processes, the initial stages of molecular interactions are dominated by the 3D shapes and shape changes of the low electron density ranges of molecular bodies. These low density ranges of the fuzzy electronic clouds are at some distance from the nuclei as well as from the imaginary bonding lines interconnecting the nuclei. Consequently, the initial interactions can be similar even for molecules containing rather different nuclei and different formal bonding patterns, as long as their peripheral electronic clouds have similar static and dynamic shape features. This is the reason why data on the shape features of 3D molecular bodies, as represented by their fuzzy electronic clouds [108,155-158], are more directly relevant to the analysis and prediction of molecular interactions than the more conventional graphs depicting the bonding patterns of molecules. The shape analysis methods based on the topological features of 3D molecular bodies provide computational tools applicable for a variety of small- and large-scale chemical and biochemical problems [43,155-158,190-199,254,262,342,345,407,408].

There is a strong tradition in research efforts towards the quantitative interpretation of chemical properties in terms of structural features of molecules (see, e.g., references [409-411]). In most physical organic chemistry studies, as well as in most synthesis design, molecular design and conventional QSAR (Quantitative Structure - Activity Relations) analysis (see, e.g., references [412-442]), the concept of structure is usually interpreted in terms of graphs depicting formal bonding patterns and structural diagrams depicting the 3D versions of these bonding patterns. In view of the special importance of 3D shape features of formal molecular bodies, an improved performance is expected from a more recent proposal on QShAR (Quantitative Shape - Activity Relations) analysis [262], based on rigorous 3D molecular shape analysis techniques. Data bases containing easily retrievable, comparable, and interpretable 3D shape data on molecular bodies appear to have an important future role in molecular engineering and drug design, and in systematic studies on the relations among functional groups in organic chemistry [262,264].

One practical aspect of the systematic shape analysis approach is of special importance for drug design and molecular engineering applications. Even though the actual relations between molecular shapes and chemical or biochemical properties may be very complex and poorly understood in some cases, for molecular design purposes a detailed understanding is not always necessary. The shape analysis of sequences of molecules of known chemical properties or known biochemical and pharmacological activities can answer the question of *what* molecular shape features are important for a given effect, even if one does not fully understand *why*. Hence, the results of shape analysis can be used directly to predict biochemical activity of new molecules. Of course, by learning better answers to the question of *what*, we are going to be able to give a more comprehensive answer to the question of *why*.

7.1   Computer Screening of Molecular Sequences by Shape Code Comparison

Most of the current approaches to the evaluation of the results of molecular modeling rely on visual inspection. Consequently, these approaches have a strong subjective component and the results are not necessarily reproducible, especially if a large number of molecular models are compared visually. The molecular Shape Group Method (SGM) [108,155-158] provides an alternative to visual evaluation of the results of molecular modeling. As it has been shown in the previous chapter, SGM and the related shape code methods are suitable for *nonvisual, algorithmic evaluation of molecular shape similarity and shape complementarity in large families of molecules.* The associated *local* shape matrix and shape code techniques are suitable for both local similarity and local shape complementarity analysis. A similarity ranking of molecular sequences, or a ranking of these sequences according to a shape complementarity measure with respect to a reference molecule such as an enzyme, helps to make predictions on the expected biochemical activity and drug potency.

The above topological shape analysis techniques can replace visual shape comparisons of molecular models on the computer screen with precise, reliable, and reproducible numerical comparisons of topological shape codes. These comparisons and the similarity or complementarity rankings of molecular sequences can be performed by the computer automatically. This eliminates the subjective element of visual shape comparisons, a particularly important concern if large sequences (e.g. several thousands) of molecules are to be compared. In the data banks of most drug companies there is information stored on literally hundreds of thousands of molecules, and their detailed shape analysis by visual comparison on a computer screen is clearly not feasible. By contrast, automatic, numerical, topological shape analysis by computer is a viable alternative.

The input data for the shape analysis methods are provided by well-established quantum chemical or empirical computational methods for the calculation of electronic charge distributions, electrostatic potentials, fused spheres Van der Waals surfaces, or protein backbones. The subsequent topological shape analysis is equally applicable to any existing molecule or to molecules which have not yet been synthesized. This is precisely where the predictive power of such shape analysis lies: based on a detailed shape analysis, a prediction can be made on the expected activity of all molecules in the sequence and these methods can select the most promising candidates from a sequence of thousands of possible molecules. The actual expensive and time-consuming synthetic work and various chemical and biochemical tests of new compounds can be focused on a few of the most promising molecules, leading to important savings in human efforts, financial resources, and reducing (but not eliminating) the need for animal and human experiments. Indeed, to the arsenal of methods of *in vitro* and *in vivo* experiments, one can add the methods of *"in computo"* techniques.

7.2   Integrated Main and Side Effect Analysis by Shape Correlations

There is an important parallel between the complexity of molecular shape and the complexity of multiple biochemical effects of molecules. Molecular shape is a multiparameter property: in order to characterize in any detail the shape of a molecule, a single number is insufficient. Even a crude description of molecular shape requires several numbers. Similarly, a molecule seldom, if ever, has just a single biochemical effect; most often, a molecule can interact in many different ways with numerous enzymes and biochemical systems. In QSAR studies the importance of side effects is well recognized (see, e.g., references [439-442]), yet in many current drug design applications of molecular modeling, one-dimensional correlations are sought between a specified type of biochemical activity and a specified aspect of molecular shape. When correlating shape properties with a given type of biochemical activity in a series of molecules, a molecular ordering is usually established based on experimentally observed activities, and the goal is to establish a parallel or nearly parallel ordering of molecules based on some shape property. If such parallel orderings are found for a set of experimentally tested molecules, then the corresponding shape features are thought to be important and are used in a predictive sense: for a new molecule of enhanced shape features a high biochemical activity is predicted. This is essentially a one-dimensional "shape calibration" approach that fails to exploit the multidimensional (multiparameter) nature of molecular shapes, and the multiple biochemical activities of molecules.

Whereas most drug design efforts up to date have been focused on the main biochemical effects of potential drug molecules, the Shape Group Method and related shape code techniques are also applicable for multiple similarity ranking based on local and overlapping molecular shape features. For a combined main effect and multiple side effect analysis, several different shape orderings can be generated, based on the information of a few molecules of experimentally known side effects. These multiple shape orderings can be defined by the shapes of functional groups, by individual local shape motifs, or by combinations of shape motifs which enhance a particular side effect. This approach provides the basis for a *systematic analysis of both main effects and of several potential side effects in large sequences of potential drug molecules.*

If sufficient input data, experimental or theoretical information on the main biochemical effect and also on various side effects are available for a series of molecules, then these molecules can serve as the basis for a detailed, multiparameter (multidimensional) shape-biochemical property analysis.

In a computational sense, the problem is converted into a multidimensional discrete shape optimization problem in a space spanned by the various orderings of the molecular sequence, according to the main and side effects. If the total number of experimentally measured biochemical properties (the main and side effects) is k, then, based on these measured properties, the molecular shapes in the given molecular series can be assigned to various points of a formal k-dimensional parameter space. A multidimensional (multiparameter) distribution of shape codes for main and multiple side effect QShAR analysis is the basic tool for the study of integrated main and side effect correlations. The distribution, clustering, and various patterns of the arrangements and interrelations of shape codes in the parameter space provide a detailed multiparameter correlation between shape and multiple biochemical effects. Such multidimensional analysis gives more information than a simple combination of individual single effect correlations. Since the entire family of biochemical activities is treated in an integrated manner, the predictive value of the approach is enhanced.

7.3   Shape-Driven Molecular Design and Molecular Engineering

All properties of molecules and molecular aggregates, industrial materials, agricultural chemicals, and drugs for medicine are influenced by the shapes of their atomic arrangements and their shape changes in chemical processes. These shapes are determined by two main components: the arrangements of atomic nuclei and the arrangements of the electronic charge clouds, generating a formal molecular body. Any chemical or biochemical process (reaction) is, in fact, a rearrangement of these shapes. Hence, shape analysis of three-dimensional bodies of electron distributions is of fundamental importance in systematic approaches towards actually designing new molecules, novel drugs, and industrial materials. This realization has been a motivating factor for the development of local convexity and relative convexity analysis of molecular surfaces leading to the Shape Group Methods [155-158,199,254], as well as for efficient numerical implementations of local convexity analysis developed for large molecules [443].

Molecular shapes and shape changes determine chemical properties. Even without knowing how a given shape feature leads to a particular chemical property, if a correlation is found, then the mere presence of the shape feature can already be used to suggest the corresponding chemical property. For the computer based design of new molecules, drugs, and industrial materials, it is usually sufficient to know what are the important shape features, and by combining these features in computer designed molecular models, one is able to propose new molecular structures for specific purposes.

In multiple shape comparisons, efficient, algorithmic shape analysis methods are of particular importance. The shape group and shape code methods provide a framework for such analysis; however, the input information they require, such as the 3D electron densities or electrostatic potentials often involve time consuming calculations. This is the case for large molecules or molecular systems, important in drug design and molecular engineering applications. Efficient calculation and representation of these molecular functions is of special importance in such cases.

Dot representations of formal molecular surfaces are useful tools for the visualization of molecular shapes [87,177]. Most of these methods use either standard atomic radii and fused sphere representations [86] or a few discrete density or potential values for displaying a selection of isoproperty surfaces. It is advantageous to augment these approaches with methods capable of a continuous variation of dot representations of molecular contours. Such a technique is useful for modeling the change of a MIDCO G(a) if the contour density value a continuously changes from a high value to a value near zero.

Below we shall describe a technique proposed [43] for the generation of a simple density scalable ("inflatable") dot representation of MIDCO's.

The Fused Sphere Guided Homotopy method (FSGH) has been designed for continuous transformations between dot representations of different isodensity surfaces of a given molecule [43]. This is a simple and rather fast computational method for representing families of isodensity contour surfaces of large molecules, for any desired density value. The FSGH method gives a faithful approximation for a whole family of MIDCO surfaces. The main idea of the method is based on the observation that actual MIDCO's of small (e.g., 0.002 a.u.) electron density values are well approximated by fused sphere model surfaces using the customary values of atomic Van der Waals (VDW) radii [86]. By selecting a quasi-uniform point distribution on each sphere and by scaling the radii of the atomic spheres, a linear charge density interpolation can be carried out between points obtained from one another by the scaling. By repeating this interpolation for all such point pairs for a selected charge density value a, an easily computable dot representation of the molecular isodensity surface is obtained, approximating the MIDCO G(a) of the corresponding density threshold value a. Since the scaling of the radii and the interpolation correspond to homotopy transformations, whereas the interpolation of isodensity surfaces is guided by the fused spheres, the technique is called the Fused Sphere Guided Homotopy (FSGH) method [43].

Fortuitously, for most molecules, the MIDCO's G(a) of the chemically most important small density threshold values a are those where the deviations are small from the simple fused sphere model surfaces. The usual Van der Waals surfaces fall within this range. For a molecule containing N nuclei, these VDWS's are obtained as the envelope surfaces of N interpenetrating spheres

$$S_1, S_2, \ldots, S_N, \tag{7.1}$$

of suitably chosen radii

$$r_1, r_2, \ldots, r_N. \tag{7.2}$$

The spheres are centered on the N atomic nuclei of position vectors

$$\mathbf{r}_1, \mathbf{r}_2, \ldots, \mathbf{r}_N, \tag{7.3}$$

respectively.

The close resemblance between such fused sphere models and MIDCO's is valid not only for a specified density threshold value a, such as a = 0.002, but for

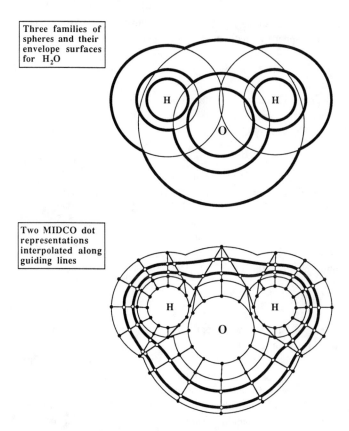

Three families of spheres and their envelope surfaces for H₂O

Two MIDCO dot representations interpolated along guiding lines

Figure   7.1   Illustration of the principle of the Fused Spheres Guided Homotopy Method (FSGH), applied for the generation of dot representations of density scalable MIDCO surfaces for the water molecule. Three families of atomic spheres (thin lines) and their envelope surfaces (heavy lines) are shown in the upper part of the figure. In the lower part of the figure, the selected point sets on the innermost family of spheres are connected by interpolating lines to the exposed points (black dots) on the envelope surfaces of two enlarged families of spheres. Linear interpolation along the lines for two selected density values leads to two families of white dots, generating approximations of two MIDCO's (heavy lines in the lower figure).

a wide range of density threshold values. This can be exploited in generating point sets for simple computer representations of MIDCO's, and for representing transformations between such contour surfaces $G(a)$ and $G(a')$ if the density value of the contour changes from a to a'.

An illustration of the principle of the Fused Spheres Guided Homotopy Method (FSGH) for the generation of dot representations of density scalable MIDCO surfaces is shown in Figure 7.1. The starting model of the FSGH dot representations is the Van der Waals surface (VDWS). We assume that the location of the center of each sphere from which the VDWS is generated is fixed, as given by Equation (7.3). A suitably small, common constant factor $k < 1$ is applied to the radii of each sphere so that, with the new radii

$$p^0{}_1 = kr_1, \quad p^0{}_2 = kr_2, \dots, p^0{}_N = kr_N, \tag{7.4}$$

no two of the new spheres have common points. These new spheres, denoted as

$$S^0{}_1, S^0{}_2, \dots, S^0{}_i, \dots, S^0{}_N, \tag{7.5}$$

are collectively referred to as the set of smallest spheres or the set of spheres of serial index $t=0$, where t is indicated as a superscript.

Based on these spheres, n new sets of spheres are generated by applying a series of increasing scaling factors $k_1, k_2, \dots, k_t, \dots, k_n$,

$$1 < k_1 < k_2 < \dots < k_t < \dots < k_n \tag{7.6}$$

for the radii $p^0{}_j$ of Equation (7.4). The resulting envelope surfaces

$$F_1, F_2, \dots, F_t, \dots, F_n \tag{7.7}$$

of these sets of spheres will approximate a sequence of MIDCO's $G(a)$ of decreasing density threshold values a.

However, this sequence of envelope surfaces of gradually enlarged fused spheres will not, in general, approximate the MIDCO's adequately for some practical applications; in particular, at the seams of interpenetrating spheres this representation does not follow the corresponding MIDCO $G(a)$ well.

A different approximation is obtained by an interpolation method that uses selected points on subsequent spherical surfaces to define lines along which one-dimensional interpolations are carried out for charge density. For this purpose, on each sphere $S^0{}_i$ of the set $t=0$, a set of m points,

$$P^0{}_i = \{r^0{}_{i1}, r^0{}_{i2}, \dots, r^0{}_{im}\} \tag{7.8}$$

is chosen from a quasi-uniform distribution along the spherical surface. The interpolation uses the point images that the enlargement process of the spheres creates from these points $\{r^0{}_{i1}, r^0{}_{i2}, \dots, r^0{}_{im}\}$ selected on the set of smallest spheres. That is, in each step t of the enlargement process by the scaling factor k_t, a new set of points

$$P^t{}_i = \{r^t{}_{i1}, r^t{}_{i2}, \dots, r^t{}_{im}\} \tag{7.9}$$

is generated on each new sphere

$$S^t{}_1, S^t{}_2, \dots, S^t{}_i, \dots, S^t{}_N. \tag{7.10}$$

These points are images the enlargement (scaling by k_t) creates from the surface points of sets $P^0_i = \{r^0_{i1}, r^0_{i2}, \ldots, r^0_{im}\}$ on the smallest spheres. We refer to the members of set P^t_i as the selected points on the t-th set of spheres.

Let us denote by

$$P^t_F = \{r^t_1, r^t_2, \ldots, r^t_u, \ldots, r^t_{vt}\} \tag{7.11}$$

the set of all those v_t surface points of all P^t_i sets in step t which fall on the envelope surface F_t. This is the subset of all the selected points on all the spheres of step t that are not buried inside F_t. For the initial disjoint spheres of family t=0, all the selected points are on the surface of the envelope surface F_0 of the spheres; this envelope is, in fact, the collection of the spheres. Hence, P^0_F is the union of all P^0_i sets.

The point pair r^t_{ij} and r^{t+1}_{ij} on two subsequent envelope surfaces F_t and F_{t+1}, respectively, is referred to as a *pair by enlargement*, if point r^{t+1}_{ij} is the image of point r^t_{ij} when the sphere S^t_i is scaled by the factor k_{t+1} / k_t.

In order to generate a scheme for density interpolation, the points occurring in subsequent sets P^t_F and P^{t+1}_F on subsequent envelope contours F_t and F_{t+1}, respectively, are interconnected according to the following criterion:

Line generation criterion. Each point $r^{t+1}_{u'}$ of set P^{t+1}_F of envelope F_{t+1} is connected by a straight line segment $s(r^t_u, r^{t+1}_{u'})$ to its unique pair by enlargement, r^t_u, if the point r^t_u falls on the previous envelope surface F_t. Otherwise, each remaining point of set P^t_F or set P^{t+1}_F on either envelope surface, not yet assigned a pair by the above condition, is connected by a straight line segment $s(r^t_u, r^{t+1}_{u'})$ to the *nearest* point of the point set P^{t+1}_F or P^t_F, respectively, of the *other* envelope surface.

If the above rule is applied to the entire sequence of envelope surfaces, then one obtains a collection of at most $(n \times m \times N)$ line segments generating a set of piecewise linear broken lines of approximately uniform distribution and approximately perpendicular to the tangent plane of each local spherical domain where the lines cross the envelope surfaces. Some of these lines may merge and possibly separate again as some of the selected points get buried and possibly reappear in the enlargement process of the spheres.

Each of these lines is composed of straight line segments $s(r^t_u, r^{t+1}_{u'})$, connecting subsequent envelope surfaces. We assume that function values of the electronic charge density $\rho(r)$ are available at each r^t_u point of each envelope surface F_t. One can use linear interpolation between the endpoints r^t_u and $r^{t+1}_{u'}$ of each line segment $s(r^t_u, r^{t+1}_{u'})$ in order to find points with a desired electronic charge density value a.

The points located by this procedure along the line segments generate a dot representation of the electronic isodensity contour (MIDCO) surface G(a) for any value a of the contour density, as long as $a_m \leq a \leq a_0$, where a_m is the *maximum* density value for points along the *low density* envelope F_n in the union of all the P^n_i sets, and a_0 is the *minimum* density value for points along the *high density* envelope F_0 in the union of all the P^0_i sets.

The task is to locate all the points $\mathbf{r}(a)$ on each of the generated line segments $s(\mathbf{r}^t_u, \mathbf{r}^{t+1}_{u'})$ that fulfill the following conditions:

$$\mathbf{r}(a) \in s(\mathbf{r}^t_u, \mathbf{r}^{t+1}_{u'}), \qquad\qquad\qquad (7.12)$$

and

$$\rho(\mathbf{r}(a)) = a . \qquad\qquad\qquad (7.13)$$

If such a point exists along the line segment $s(\mathbf{r}^t_u, \mathbf{r}^{t+1}_{u'})$, then it can be located by linear interpolation. In practice, it is sufficient to attempt interpolations for those line segments which fulfill either one of the following two conditions:

$$\rho(\mathbf{r}^t_u) \leq a \leq \rho(\mathbf{r}^{t+1}_{u'}), \qquad\qquad\qquad (7.14)$$

or

$$\rho(\mathbf{r}^t_u) > a > \rho(\mathbf{r}^{t+1}_{u'}). \qquad\qquad\qquad (7.15)$$

The resulting set $P(a)$ of all interpolated points $\mathbf{r}(a)$ can be regarded as a dot representation of the MIDCO surface $G(a)$, a representation useful in drug design, molecular engineering, and some more general computer graphics applications of molecular modeling. However, the above representation has some further advantages: along each of the $(m \times N)$ lines a simple, analytic functional form is available for the approximation of electronic density. This allows one to explicitly construct a relatively simple function for an approximate representation of the *continuous deformations* of MIDCO surfaces $G(a)$, regarded as functions of the contour density value a. Such continuous deformations, *homotopies,* are advantageous in the analysis of the dependence of molecular shape properties on electronic charge density. The name of the method, Fused Spheres Guided Homotopy Method (FSGH), reflects the fact that the homotopy transformation is obtained by following the changes in envelope surfaces of families of fused spheres of gradually increasing sizes [43].

Note that the same FSGH technique is applicable to many other molecular functions which can be approximated by contour surfaces not drastically different from a fused sphere VDWS. This condition is not in general valid for electrostatic potentials since MEP surfaces may show even sign changes in various regions of the space. However, the method is applicable for local regions of the electrostatic potential (e.g., one may apply the FSGH technique separately for some positive and negative domains).

The FSGH representations provide a practical tool for the implementation of a variety of techniques suggested for molecular shape analysis. In particular, the FSGH method is applicable to approximate the entire charge density clouds and to generate the complete topological shape codes for large molecules, using semi-empirical or empirical charge density functions, even if for such large systems direct, high quality quantum chemical electron density calculations are not feasible.

The related recent development of Density Scalable Atomic Spheres, DSAS [255] allows one to build VDWS-like fused sphere approximations of MIDCO's for any density threshold value a (with the exception of very high densities such as those in the immediate vicinity of the nuclei). The basis of this technique is a family of scaling functions [255] developed for generating radii $r_A(a)$ of formal atomic spheres along which the electronic density of atom A is any selected constant value a, within a chemically important range of electronic densities. These "density scalable" radius functions $r_A(a)$ have been determined [255] for all the atoms A commonly encountered in molecular modeling problems.

The technique of density scalable atomic spheres provides an inexpensive approximation for MIDCO's of large molecules, such as those studied in typical drug design and molecular engineering problems, if direct quantum chemical density calculations are not feasible. Furthermore, the density scalable radii also serve as natural starting points for the construction of the families of spheres required for the FSGH technique described above.

Isopotential contours of the composite nuclear potentials (NUPCO's, see Chapter 4), provide an inexpensive, approximate shape representation that can be computed easily even for very large molecules. Although NUPCO's only approximate the MIDCO's of molecules, the family of NUPCO's of a molecule describes an important molecular property that has a major effect on the actual molecular shape. Consequently, NUPCO's can be used for direct comparisons between molecules, and similar NUPCO's are likely to be associated with similar molecular shapes. All the shape analysis techniques originally developed for MIDCO's are equally applicable to NUPCO's. The shape groups, the (a,b) parameter maps [where a is the nuclear potential threshold of a NUPCO G(a)], the shape matrices, shape codes, and the shape globe invariance maps of NUPCO's of molecules can serve as inexpensive methods for the detection and evaluation of a particular aspect of molecular similarity.

7.4   A Summary of the Main Components of the   QShAR   Approach

The QShAR (Quantitative Shape-Activity Relations) method, combined with the integrated main and side effect modeling of bioactive molecules, forms the conceptual basis of the approaches described in this chapter. The density scalable FSGH method for a simple representation of molecular bodies, in combination with the Shape Group Method and various other shape code approaches for quantitative shape analysis, as well as the multiple shape ranking methods for integrated main and side effect analysis, are the components of a computational implementation of the basic concepts.

The resulting framework is a computer modeling approach for a systematic treatment of multiple biochemical effects in drug design, herbicide, fungicide, and pesticide design, preventive environmental toxicology, and shape-driven molecular engineering. These methods can serve as tools in the construction of new supramolecular materials with prescribed shape-dependent chemical and biochemical properties.

The list below is a brief summary of the basic theoretical and the main methodological components of the QShAR approach.

1. The GSTE Principle: Geometrical Similarity as Topological Equivalence is a general principle for similarity analysis [108].

2. The method of (P,W)-similarity assessment [108] is a general scheme for the quantification of molecular similarity in terms of shape representations P (e.g., electronic charge isodensity surface) and shape descriptor W (e.g., molecular shape groups).

3. SGM: the Shape Group Methods [155-158,199]. The original shape groups are the homology groups of molecular contour surfaces, truncated according to local curvature criteria or according to ranges of values of other molecular functions (e.g., electrostatic potential) mapped on an isodensity contour (MIDCO) surface. The shape groups are the basic tools for converting 3D shape information, through a topological filter, to numerical shape codes [43,109,196,351].

4. Shape codes [43,109,196,351,408]. The simplest topological shape codes derived from the shape group approach are the (a,b) parameter maps, where a is the isodensity contour value and b is a reference curvature against which the molecular contour surface is compared. Alternative shape codes and local shape codes are derived from shape matrices and the Density Domain Approach to functional groups [262], as well as from Shape Globe Invariance Maps (SGIM).

5. Dynamic shape description of nonrigid molecules [107,158,197,199,408] is accomplished by the calculation of shape code preserving conformational domains of flexible molecules, leading to Dynamic Shape Codes. Similar dynamic shape codes are obtained from Shape Globe Invariance Maps (SGIM) for global shape analysis: this method is applicable for a complete mapping of all topologically different views of a molecule enclosed within a sphere, while moving the observer along the sphere. Shape Globe Invariance Maps (SGIM) are applicable for nonvisual characterization of dynamic shape properties of protein folding.

6. Resolution Based Similarity Measures (RBSM). These methods are based on evaluating the level of resolution required to distinguish molecules; the higher the required level of resolution, the more similar are the molecules [240,243, see also 54,55].

7. The FSGH method (Fused Sphere Guided Homotopy method) [43]. This method has been designed for the construction of approximate, density scalable ("inflatable") isodensity contour surfaces and their dot representations (i.e., for continuous transformations between different isodensity surfaces of a given molecule).

8. Density Scalable Atomic Sphere (DSAS) surfaces [255]. This technique generates radii for atomic spheres for any desired electron density at the surface. The method is used for inexpensive representations of MIDCO's of large molecules, in combination with the Fused Sphere Guided Homotopy method (FSGH) [43].

9. The integrated main effect and multiple side-effect analysis by multiple shape ranking of bioactive molecules, based on the density scalable FSGH and Shape Group Methods, are suitable for shape analysis of large molecules.

10. Multiple shape ranking of protein structures, based on local FSGH analysis of active sites, is combined with global shape analysis of folding patterns using a version of the SGIM (Shape Globe Invariance Map) method. This latter method generates a map of all topologically distinguishable views of the folded chain, represented by a space curve. The spherical map, now analogous to a molecular contour surface, can also be analyzed by the Shape Group Method, hence a common methodology can be applied for both problems.

11. Isopotential contours of the composite nuclear potentials (NUPCO's), provide an inexpensive approximate shape representation that can be computed easily even for very large molecules.

CHAPTER

8

SPECIAL TOPICS:
SYMMETRY AND APPROXIMATE SYMMETRY,
SYMMETRY DEFICIENCY MEASURES,
SYNTOPY, AND SYMMORPHY

This chapter presents a brief review of some special aspects of the topological theory of molecular shape, with emphasis on various treatments of approximate symmetry. If a molecular arrangement has some nontrivial symmetry, then this symmetry can be exploited for the simplification of the study and prediction of many physical and chemical properties of the molecule. In many cases, however, the molecular arrangements may deviate from their ideal symmetry, yet many molecular properties remain similar to those present in the symmetric arrangement. Approximate symmetry, symmetry deficiency, the methods for their quantification, and various algebraic treatments of approximate symmetries preserving some of the features of the standard group theoretical description of point symmetry, are the subject of this chapter.

8.1   Symmetry and Imperfect Symmetry

Consider two molecular arrangements, K_1 and K_2, both of C_2 nuclear point symmetry, and both derived from the same equilibrium molecular structure K_0 of C_{2v} point symmetry. Let us assume that arrangement K_1 differs only slightly from K_0, whereas K_2 is very different from K_0. Furthermore, we assume that it takes a large distortion of K_2 to convert it into any arrangement K that has C_{2v}

point symmetry. Intuitively, one may regard K_1 of C_2 symmetry as having "almost" C_{2v} point symmetry, whereas K_2, also of C_2 symmetry, has little resemblance to any arrangement of C_{2v} point symmetry. The two arrangements, nuclear configurations K_1 and K_2 of the same point symmetry, show different levels of symmetry resemblance to C_{2v} point symmetry, in other words, they exhibit *different levels of imperfect* C_{2v} *symmetry*.

A rather simple measure of such symmetry imperfection has been defined [345] as the minimum distortion required for converting a given molecular arrangement into one with the specified point symmetry, where the distortion is measured by the distance in M space, a metric nuclear configuration space [106]. Here we shall use the configuration space formalism of the original approach [345], where d represents the distance in the metric configuration space M [106]. As discussed in Chapter 2, if K_1 and K_2 are two conformations of a molecule, or two arrangements of two different molecules with the same stoichiometry, then their M space distance $d(K_1, K_2)$ is a well-defined numerical measure for the degree of their dissimilarity. As described in [345], using the above distance as a dissimilarity measure [106], one can answer questions of the following type [345]:

1. how different is an actual molecular arrangement K_1 from another arrangement K_2 ?
2. how much is the deviation of an actual configuration K_1 from a family F of molecular arrangements?

For question 1 the distance $d(K_1, K_2)$ provides the answer, whereas for questions of type 2 the following minimum distance

$$d(K_1, F) = min \{ d(K_1, K_2) : K_2 \in F\} \qquad (8.1)$$

can be used to define an appropriate measure of deviation [345].

In our actual problem, the family F of special arrangements is the set G of all molecular arrangements K of the given stoichiometry with the specified point symmetry group g. The distance

$$d(K_1, G) = min \{ d(K_1, K_2) : K_2 \in G\} \qquad (8.2)$$

defines a mathematically proper measure of the deviation of configuration K_1 from symmetry G. Consequently, $d(K_1,G)$ is a well-defined symmetry deficiency measure.

In another application of the same method, the quantity

$$d(K_1, A) = min \{ d(K_1, K_2) : K_2 \in A\} \qquad (8.3)$$

is a measure of chirality, where family A contains all achiral molecular arrangements K of the given stoichiometry.

Note that the distance function d of the nuclear configuration space M involves a mass-scaling of nuclear coordinates [106], hence d incorporates a "natural rescaling" of differences between various nuclear arrangements. Consider two different distortions of a given molecular arrangement K, one involving the displacement of a light nucleus, the other a displacement of a heavy nucleus. For *equal* displacements in the ordinary 3D space, the displacement of the heavy nucleus is more significant. Due to the mass-scaling of coordinates, the displacement of the heavy nucleus also leads to a greater d distance in the nuclear configuration space M [106].

Imperfect symmetry can be viewed in the broader context of molecular distortions. In this book the emphasis is on descriptions and measures based on topological arguments [54,55,106,108,158,240,243,247,248,345,444,445], or on the related concepts of syntopy [252,394,395] and symmorphy [43,108] discussed briefly in subsequent parts of this chapter. For a variety of alternative approaches and important additional insight, the reader may consult references [58,446-449].

**8.2   The Quantification of Approximate Symmetry:
        Symmetry Deficiency Measures**

Chirality can be regarded as the lack of certain symmetry elements, and chirality measures are in fact measures of symmetry deficiency. In the case of three-dimensional chirality, the lacking point symmetry elements are reflection planes σ and rotation-reflections S_{2k} of even indices. Note, however, that the lacking symmetry elements can be of different nature in different dimensions, and at the end of this section, a proof will be given for an elementary result [240] on the dimension-dependence of chirality.

By analogy with chirality and various chirality measures, more general symmetry deficiencies and various measures for such symmetry deficiencies can be defined with reference to an arbitrary collection of point symmetry elements. We shall discuss in some detail only the three-dimensional cases of symmetry deficiencies, however, as it has been pointed out in reference [240], all the concepts, definitions, and procedures listed have straightforward generalizations for any finite dimension n.

Following the approach of the original description [240], we shall first consider chirality. For a chiral object T, the largest achiral object that fits within T, as well as the smallest achiral object that contains T are of special importance. We shall compare the volumes v of these objects, and use these comparisons to assess the degree of the deviation of the object T from achirality.

The above is a simple idea; however, as the following example shows, some caution is in order. Consider two solid balls of the same radius, where one ball has a spiral line issued from its surface. The first object is achiral whereas the second one is chiral, and the first object is the largest achiral object that fits within the second one. Clearly, the two objects have the same volume, hence comparing volumes is not appropriate for assessing the degree of chirality (i.e., the degree of "achirality deficiency") of the second object. This problem is caused by the presence of the infinitely thin spiral line of zero volume. In order to avoid such pathological cases, we shall consider only objects T that are "nowhere infinitely thin" and have finite, nonzero volume [240].

Following the original framework described in detail in reference [240], we list some of the relevant concepts and definitions below. The objects defined lend themselves naturally for various measures of chirality and in a more general sense, for measures of more general types of symmetry deficiency.

Set M' is a *maximal achiral subset* of T if M' is achiral, $M' \subset T$ and if no achiral set M" exists such that $M' \subset M"$, $M' \neq M"$, and $M" \subset T$. Note that such a subset M' is not necessarily unique for a given set T.

Set M is a *maximal volume achiral subset* of T if M is achiral, $M \subset T$ and if for all maximal achiral subsets M' of T, $v(M') \leq v(M)$. Note that, for a given set T, such a subset M is not necessarily unique either, however, the volume $v(M)$ is a unique number for each T.

Set N' is a *minimal achiral superset* of T if N' is achiral, $T \subset N'$ and if no achiral set N" exists such that $N" \subset N'$, $N' \neq N"$, and $T \subset N"$. Note that such a subset N' is not necessarily unique for a given set T.

Set N is a *minimal volume achiral superset* of T if N is achiral, $T \subset N$ and for all minimal achiral supersets N' of T, $v(N) \leq v(N')$. For a given set T, such a set N is not necessarily unique either, however, the volume $v(N)$ is a unique number for each T.

If T is achiral then both M and N are unique and $M = N = T$.

The expressions

$$\chi_M(T) = 1 - v(M)/v(T) \tag{8.4}$$

and

$$\chi_N(T) = 1 - v(T)/v(N) \tag{8.5}$$

define two chirality measures, where the first measure, $\chi_M(T)$, agrees with the measure obtained using the maximum overlap criterion between mirror images [46,48-53,58,242].

The actual determination of a set M for some chiral set T and the calculation of the volume $v(T)$ are usually rather difficult problems (see some relevant comments in references [51-53,58,240,242]), and the same applies for superset N. However, within a RBSM framework, the analogous chirality measures given in terms of a discretization procedure using polycubes (or lattice animals in 2D) [240] do not require the explicit determination of a maximal volume (area) achiral subset M and the calculation of its exact volume (or area) $v(M)$.

Recalling that chirality is just a special case of symmetry deficiency, the above concepts and ideas are applicable for more general symmetry deficiencies. Consider a family

$$\mathbf{R} = \{R_1, R_2, \ldots, R_m\} \tag{8.6}$$

of point symmetry elements. We say that set U is an **R**-*set* if U has all point symmetry elements of family **R**. Set V is an **R**-*deficient set* if V has none of the point symmetry elements of family **R**. Note, however, that it takes only infinitesimal distortions to lose a given point symmetry element, hence, unless further restrictions are applied, the volume difference between a set of a specified point symmetry and another that does not have this symmetry can be infinitesimal. Consequently, **R**-deficient subsets and **R**-deficient supersets of an **R**-set can be almost identical. Nevertheless, the definitions involving symmetry deficient sets lead to nontrivial results if the sets considered can differ only by fixed, positive volume increments, such as is the case for polycubes. The actual symmetry deficiencies of more general continuum sets, such as formal molecular bodies enclosed by MIDCO surfaces, are defined in terms of deviations from maximal **R**-subsets and minimal **R**-supersets.

Set M' is a *maximal **R**-subset* of T if M' is an **R**-set, M'⊂ T and if no **R**-set M" exists such that M'⊂M", M' ≠ M", and M"⊂T. Note that M' is not necessarily unique for a given set T.

Set M is a *maximal volume **R**-subset* of T if M is an **R**-set, M⊂T, and if for all maximal **R**-subsets M' of T, v(M') ≤ v(M). Set M is not necessarily unique for a given set T; however, the volume v(M) is already a unique number for each T.

Set N' is a *minimal **R**-superset* of T if N' is an **R**-set, T⊂N', and if no **R**-set N" exists such that N"⊂N', N' ≠ N", and T⊂ N". Set N' is not necessarily unique for a given set T.

Set N is a *minimal volume **R**-superset* of T if N is an **R**-set, T⊂ N, and if for all minimal **R**-supersets N' of T, v(N) ≤ v(N'). A set N is not necessarily unique for a given set T; however, the volume v(N) is a unique number for each T.

If T is an **R**-set then both M and N are unique and M = N = T.

Set M' is a *maximal **R**-deficient subset* of T if M' is an **R**-deficient set, M'⊂T, and if no **R**-deficient set M" exists such that M'⊂ M", M' ≠ M", and M"⊂ T. Set M' is not necessarily unique for a given set T.

Set M is a *maximal volume **R**-deficient subset* of T if M is an **R**-deficient set, M⊂T, and if for all maximal **R**-deficient subsets M' of T, the relation v(M') ≤ v(M) holds. Set M is not necessarily unique for a given set T; however, the volume v(M) is a unique number for each T.

Set N' is a *minimal **R**-deficient superset* of T if N' is an **R**-deficient set, T⊂N', and if no **R**-deficient set N" exists such that N"⊂ N', N' ≠ N", and T⊂N". Set N' is not necessarily unique for a given set T.

Set N is a *minimal volume **R**-deficient superset* of T if N is an **R**-deficient set, T⊂ N, and if for all minimal **R**-deficient supersets N' of T, the relation v(N) ≤ v(N') holds. Set N is not necessarily unique for a given set T; however, the volume v(N) is a unique number for each T.

If T is an **R**-deficient set then both M and N are unique and M = N = T.

If M is a maximal volume **R**-subset of T, and N is a minimal volume **R**-superset of T, then the expressions

$$\delta_{R,M}(T) = 1 - v(M)/v(T) \qquad (8.7)$$

and

$$\delta_{R,N}(T) = 1 - v(T)/v(N) \tag{8.8}$$

define two **R**-deficiency measures, the *internal* **R**-*deficiency measure* $\delta_{R,M}(T)$, and the *external* **R**-*deficiency measure* $\delta_{R,N}(T)$, respectively. Their average defines the **R**-*deficiency measure*

$$\delta_{R,M}(T) = (\delta_{R,M}(T) + \delta_{R,N}(T)) / 2. \tag{8.9}$$

For any **R**-subset M', **R**-superset N', maximal volume **R**-subset M, and minimal volume **R**-superset N of any set T the relations

$$v(M') \leq v(N') \tag{8.10}$$

and

$$v(N) - v(M) \leq v(N') - v(M') \tag{8.11}$$

hold.

Evidently, if the family **R** contains a symmetry element of reflection σ or one of the rotation-reflections S_{2k}, then the **R**-sets are achiral sets. For any set T (chiral or achiral), the various extremal achiral sets can be generated by special **R**-sets which are extremal over all choices of families **R** containing at least one of the above point symmetry elements of σ or S_{2k}. Following the notation of reference [240], subscripts α and R are used to distinguish achiral sets and **R**-sets.

If M_α, M_R, N_α, and N_R are maximal volume achiral subset, maximal volume **R**-subset, minimal volume achiral superset, and minimal volume **R**-superset of a set T, respectively, then

$$v(M_\alpha) = \max_R \{ v(M_R) : M_R \subset T, \ \sigma \in R \ \text{or} \ S_{2k} \in R \ \text{for} \ k{>}0\} \tag{8.12}$$

and

$$v(N_\alpha) = \min_R \{ v(N_R) : T \subset N_R, \ \sigma \in R \ \text{or} \ S_{2k} \in R \ \text{for} \ k{>}0\}. \tag{8.13}$$

All the above concepts and considerations of symmetry deficiency and chirality have straightforward generalizations for any finite dimension n. Note, however, that the lack of different families of symmetry elements causes chirality in different dimensions, and chirality is obviously dimension dependent. If a given object is achiral when embedded in a space of n-dimensions, it may be chiral if embedded in a space of some different dimensions. Below we shall describe a related elementary result in more precise terms.

Following the proof given in reference [240], we shall use the standard notations: E^{n+1} denotes an (n+1)-dimensional Euclidean space and E^n is an n-dimensional subspace of E^{n+1}. If we refer to the n-dimensional chirality of an object A, then we consider its embedding in an Euclidean space E^n and reflections as well as all motions are restricted to this space.

As it has been shown in reference [240], the following holds:

Any object A *that is chiral in n-dimensions is achiral in* (n+1)-*dimensions and in any higher dimensions. Chirality may occur only in the lowest dimension where* A *is embeddable.*

Proof: The object A is chiral in n-dimensions (i.e., when embedded in space E^n). Let us denote the mirror image of A by $A^\Diamond$ and the corresponding mirror image of point $p \in A$ by $p^\Diamond$. By translations and rotation, we can always arrange A and $A^\Diamond$ in E^n so that for all their point pairs p and $p^\Diamond$ their coordinates fulfill the relations

$$p^\Diamond{}_1 = -p_1 \qquad\qquad\qquad\qquad\qquad\qquad\qquad (8.14)$$

and

$$p^\Diamond{}_i = p_i \qquad (i = 2,3, ..., n) . \qquad\qquad\qquad\qquad (8.15)$$

For this arrangement, the (n-1)-dimensional reflection hyperplane E^{n-1} in E^n is defined by

$$x_1 = 0 \qquad\qquad\qquad\qquad\qquad\qquad\qquad\qquad (8.16)$$

where x_1 is the first coordinate of a point $x \in E^n$.

Consider now the same arrangement of A and $A^\Diamond$ embedded in E^{n+1}, by regarding E^n as a subspace of E^{n+1}. A two-dimensional rotation in E^{n+1} is defined by its (n-1)-dimensional axis and by the angle α of rotation in the remaining two dimensions. [Note that in a k-dimensional space, the axis of rotation is (k-2)-dimensional.] Choose the rotation axis in E^{n+1} as the (n-1)-dimensional subset defined as the reflection hyperplane E^{n-1} of condition $x_1 = 0$ in E^n. With respect to this axis, a rotation of angle $\alpha = \pi$ in the two-dimensional plane spanned by coordinates (x_1, x_{n+1}) superimposes A on $A^\Diamond$ in (n+1)-dimensions. Consequently, the object A is *achiral* in (n+1)-dimensions (i.e., when embedded in space E^{n+1}). Furthermore, the superimposition of mirror images performed in E^{n+1} is a possible motion in any Euclidean space E^{n+k}, k>1, of which E^{n+1} is a subspace, hence A is achiral in any higher dimensions. Consequently, chirality may occur only in the lowest dimension where A is embeddable. Q.E.D.

8.3   Syntopy and Syntopy Groups

As discussed in Section 6.7, fuzzy sets are tools for treating classification problems with imprecisely defined classification criteria [382-385]. We recall that for an ordinary set A, a point p either does or does not belong to the set A. This can be restated in terms of membership functions: the membership function of the point p in set A is either one or zero. For points of ordinary sets, "belonging" is a discrete concept, resulting in a binary membership function. In fuzzy set theory, the concept of belonging is further qualified, resulting in the continuous concept of the "grade of belonging" to a given fuzzy set; the fuzzy membership function of a point can take any value from the [0,1] interval.

Imperfect symmetry can be regarded as fuzzy symmetry. The theory of fuzzy sets has found applications in many fields of engineering and natural sciences (see, e.g., references [386-393]), in particular, for the description of fuzzy molecular arrangements [103,106,251]. It is natural to consider fuzzy sets for a continuous extension of the discrete point symmetry concept to quasi-symmetric molecular structures [252,394,395].

There is strong motivation for such a generalization of symmetry. A nuclear configuration K of a molecular conformation either has or does not have a given point symmetry. This feature of symmetry is in contrast with most other molecular properties which vary continuously with the nuclear configuration. The discrete nature of the presence or absence of symmetry elements hinders the application of point group symmetry methods for general molecular arrangements. However, the discrete concept of point symmetry can be converted into a continuous one and extended to cover cases of "almost" symmetric or quasi-symmetric molecular structures and arrangements. In such a conversion some of the advantages of the group theoretical treatment of truly symmetric structures can be retained for most (in the extreme case for all) possible molecular arrangements.

One framework for this conversion is the *syntopy group* approach [252,394,395], where the sharply defined families of nuclear arrangements having a specified point symmetry are replaced by fuzzy sets (syntopy sets) of nuclear arrangements having some degree of symmetry resemblance to arrangements of perfect point symmetries. This replacement is carried out within a formalism that provides the syntopy sets with a group theoretical characterization, retaining some of the group theoretical, algebraic relations among the point symmetry operators of arrangements of some precise symmetry. This approach leads to the characterization of fuzzy syntopy sets by the syntopy groups [252,394,395]. In the syntopy model, the various algebraic groups of symmetry operations of individual configurations K are extended into larger groups. These groups are also provided with additional, continuous features, as a result of the replacement of the binary membership functions of ordinary sets with the continuous membership functions of the fuzzy syntopy sets.

There have been two types of syntopy proposed, one based on energetic [252,394], the other on purely geometric [395] resemblance of nuclear arrangements to those of specific point symmetries.

The original syntopy concept was based on an energetic criterion and on the formalism of fuzzy set theory, leading to energy-dependent syntopy groups of quasi-symmetric, general nuclear arrangements [252,394]. The fuzzy membership functions were defined for all possible nuclear configurations, expressing their "grade of belonging" to each possible ideal symmetry. These membership functions are parametrized by a threshold for the energy cost of converting each nuclear configuration to structures of ideal symmetry. The parametrization allowed to convert the syntopy model into the conventional point symmetry group model in a continuous manner, indicating that syntopy is, indeed, a generalization of point symmetry. Since the energy cost of a given geometric interconversion is, in general, different for each potential energy surface, a different syntopy model is obtained for each potential surface, that is, for each electronic state of the given stoichiometric family of atoms.

In the second approach [395], a common syntopy model is derived for all potential surfaces, that is, for all electronic states of all possible arrangements of the given collection of atoms. This syntopy is the underlying syntopy for the given stoichiometry, that is, the *fundamental syntopy* of the nuclear configuration space M specified for the given stoichiometry. The fundamental syntopy is independent of energetic considerations of individual potential surfaces, it does not depend on the energy cost of interconversions. The fundamental syntopy is parametrized in terms of a universal geometrical criterion that defines the fuzzy membership functions and the syntopy groups. The fundamental syntopy model provides the connection among all possible energy based syntopies generated by the various potential energy surfaces of the given family of atomic nuclei, that is, of the given stoichiometry.

A fuzzy set generalization of nuclear point symmetry in terms of the syntopy models is applicable to all nuclear arrangements. Using appropriate membership functions [252,394,395], syntopy provides a measure of symmetry resemblance of actual, general molecular arrangements to ideal, fully symmetric arrangements. The approach takes into account the quantum mechanical, nonlocalized nature of nuclei. The syntopy groups can be regarded as generalizations and continuous extensions of ordinary point symmetry groups from a restricted family of symmetric nuclear arrangements to all possible nuclear arrangements.

Within the syntopy model outlined above, the generalization of point symmetry is achieved by considering the symmetry resemblance of actual molecular arrangements to those of some ideal symmetry. In principle, any one of the similarity measures and symmetry deficiency measures discussed in the previous chapters is suitable to serve as parameter in the definition of fuzzy syntopy membership functions, leading to further generalization of syntopy.

8.4   Symmorphy and Symmorphy Groups

Within the syntopy model, the essential algebraic structure of point symmetry groups is retained (in fact, this structure is extended), and the elements of syntopy groups are derived from ordinary point symmetry operators [252,394,395]. There are, however, alternative approaches for the generalization of symmetry, where fundamentally different algebraic structures are used.

One such approach, the *symmorphy group approach* [43,108], is based on the extension of the family of point symmetry operators to a much richer family of operations *which preserve the general morphology of objects.* (Note that the term "symmorphy" is used in a different sense in the crystallography literature, with reference to the symmorphic space groups of crystallography, also called semi-direct products or split extensions in mathematics, see, e.g., reference [450]. The symmorphy model described in this chapter should not be confused with the symmorphic space groups of crystallography.)

Point symmetry groups provide at least a partial characterization of molecular shapes. This characterization can be improved considerably by extending the family of point symmetry operators to a much larger family of continuous transformations. Symmorphy is a particular extension of the point symmetry group concept of finite point sets, such as a collection of atomic nuclei, to a complete algebraic shape characterization of continua, such as a three-dimensional electron distribution of a molecule.

First we shall return to the concept of symmetry, taking a special approach that lends itself for generalization. Consider a fixed nuclear geometry K of a molecule, and the collection of all possible planes, lines and points in the 3D space. These planes define infinitely many reflections of the space onto itself, but only few, if any, of these reflections are actual symmetry operations for the given configuration K: those along planes that happen to be symmetry elements of the configuration. Reflections of the nuclear configuration with respect to any other of the planes lead to nuclear arrangements that are *distinguishable* from the original one (i.e., the original and the reflected image are not exactly superimposed). The configuration K provided a selection criterion for classifying all possible reflection planes: those which leave the appearance of K invariant, and those which do not. Similarly, among the infinitely many rotations about all the possible straight lines in the space, by all possible angles, only few, if any, are actual symmetry operations for the given configuration. Among the infinitely many points of the space only one, if any, is a point of inversion for this configuration K. All the reflections, all proper and improper rotations and all inversions, with respect to all planes, lines and points of the space, respectively, form a group G'. However, only a subgroup g'_K of this group G' is relevant to the actual configuration K; this subgroup g'_K is the point symmetry group of nuclear arrangement K. The generalization of symmetry to symmorphy can be based on the following observation: *It is the configuration K that selects these special reflections, rotations, and inversions from the set of infinitely many such operations of the space. The condition for selection is the indistinguishability of the original and transformed configurations.*

The above selection principle can be extended in two ways:

1. from the family of reflections, proper and improper rotations and inversions to a family G of more general transformations: to the set of all the possible homeomorphisms of the 3D space (i.e., to all continuous assignments of the points of the space to the points of the space, with continuous inverse transformations).

2. from a nuclear point distribution to more general 3D objects [e.g., to molecular charge density functions $\rho(\mathbf{r})$].

Among the transformations in family G one finds all the symmetry operations, but also all reflections in curved mirrors, nonlinear stretchings, and all continuous distortions of the space. Evidently, all possible homeomorphisms of the 3D space form a group G. Any two such transformations applied consecutively correspond to one such transformation (closure property); the unit element is the identity transformation; inverse exists for each transformation; and associativity is also guaranteed.

With reference to a given 3D object such as a molecular charge density function $\rho(\mathbf{r})$, one can take a selection from the transformations in family G, based on the same general condition used above in the case of point symmetry operations: *indistinguishability of the original and transformed objects.*

We make one concession: only the original object is required to have direct physical meaning. We assume that all mathematically possible transformations of the space can be performed, and that the original and transformed spaces and the original and transformed objects can be compared. For the example of a 3D molecular charge density function $\rho(\mathbf{r})$, all those transformations t of family G are selected for which

$$\rho(t\mathbf{r}) = \rho(\mathbf{r}) \tag{8.17}$$

for every point $\mathbf{r}$ of the 3D space.

If the density function $\rho(\mathbf{r})$ has some symmetry element, then the corresponding symmetry operation t is among the selected transformations. However, there are many more homeomorphic transformations t of the 3D space that leave the *appearance* of molecular charge density function $\rho(\mathbf{r})$ indistinguishable from the original. For example, all homeomorphic distortions t of the space that assign the points of each MIDCO G(a) to some points of the same MIDCO will leave the appearance of the density function $\rho(\mathbf{r})$ unchanged. These transformations are special with respect to the charge density $\rho(\mathbf{r})$; it is, in fact, the "object" $\rho(\mathbf{r})$ that selects these transformations from the very large family G of all possible homeomorphic transformations of the 3D space.

These special transformations are the *symmorphy transformations* [43,108] of the given "object", for example, of the electronic charge density function $\rho(\mathbf{r})$. The terminology is justified by the analogy with symmetry: in symmetry the *metric* properties are preserved, in *symmorphy* the *morphology,* the appearance of shape, is preserved.

In symmorphy transformations the shape of the object is invariant; in our example, the shape of the continuous charge density $\rho(\mathbf{r})$ remains the same. Nevertheless, the metric properties, such as the distance between two points may change.

A simple example, taken from reference [108], illustrates this point. Assume that the object undergoing a symmorphy transformation has a circular cross-section of radius equal to 1, where the circle is parametrized by an angle variable α, taken from the interval $0 \leq \alpha \leq 2\pi$. If a selected transformation t of the space has the effect of leaving all points of this circle on the same circle, but shifting points along the circle in a nonuniform manner, then the appearance of the circle will not change. For example, this is the case if t transforms the parameter α into the angle variable β, where

$$\beta = 2\pi\sin^2(\alpha/4). \tag{8.18}$$

If the angle variable α changes monotonically from 0 to 2π then the new, transformed angle variable β also changes monotonically from 0 to 2π, but at a rate that is slower in some and faster in some other subintervals, when compared to the rate of change of α. In this transformation t the shape remains the same, a circle, but the metric properties do change. The distance between two points may change as a result of the transformation.

For example, the distance of 0.517638... between the pair of points characterized by

$$\alpha = 0,$$

and

$$\alpha' = \pi/6 = 0.523599... ,$$

respectively, will change to the distance of 0.106996... of their transformed counterparts, characterized by the new angle values

$$\beta = 2\pi\sin^2(\alpha/4) = 0,$$

and

$$\beta' = 2\pi\sin^2(\alpha'/4) = 2\pi\sin^2(\pi/24) = 0.1070471...,$$

respectively.

The family of all *symmorphy transformations* of the given object $\rho(\mathbf{r})$ form a subgroup g_ρ of the group G of all homeomorphic transformations of the 3D space. This subgroup g_ρ is, in fact, defined by the shape properties of the 3D object, and it provides a complete characterization of its shape, in our case, the shape of the molecular charge density function $\rho(\mathbf{r})$.

However, the group g_ρ is much too complicated for practical purposes of molecular shape characterization. Fortunately, the behavior of transformations t of family g_ρ far away from the object $\rho(\mathbf{r})$ is of little importance, and one can introduce some simplifications. Let us assume that the 3D function considered [e.g., an approximate electron density function $\rho(\mathbf{r})$], becomes identically zero outside a sphere S of a sufficiently large radius. As long as two symmorphy transformations t_1 and t_2 have the same effect within this sphere, the differences between these transformations have no relevance to the shape of $\rho(\mathbf{r})$, even if they have different effects in some domains outside the sphere. All such transformations t of equivalent effects within the relevant part of the 3D space can be collected into equivalence classes. In the symmorphy approach to the analysis of molecular shape, these *classes* are taken as the actual tools of shape characterization.

A more precise formulation of the above idea is given as follows. At every point $\mathbf{r}$ where $\rho(\mathbf{r}) \neq 0$, the effects of transformations t of family g_ρ are compared to the effects of a selected transformation t_1. The effects of some transformations t agree with the effects of the selected transformation t_1. All transformations t for which

$$t \, \mathbf{r} \; = t_1 \mathbf{r} \tag{8.19}$$

for every point $\mathbf{r}$ such that

$$\rho(\mathbf{r}) \neq 0, \tag{8.20}$$

belong to the same equivalence class T_1. In general, there are infinitely many transformations in each equivalence class T. Since the transformations in these classes preserve the morphology of the object, for example, the morphology of the charge density $\rho(\mathbf{r})$, they are called the *symmorphy classes* T of the object $\rho(\mathbf{r})$.

If two transformations t_1 and t_2 are related to each another as t is related to t_1 in conditions (8.19) and (8.20), then t_1 and t_2 are said to be *symmorphy equivalent,* or in short, *symmorphic* to one another with respect to the given object $\rho(\mathbf{r})$. Symmorphy equivalence is denoted by

$$t_1 \;\; \boldsymbol{m}_\rho \;\; t_2, \tag{8.21}$$

and as the notation $\boldsymbol{m}_\rho$ indicates, it is dependent on the object $\rho(\mathbf{r})$. The symmorphy relation $\boldsymbol{m}_\rho$ is reflexive, symmetric, and transitive, hence it fulfills the conditions for an equivalence relation.

The family of all symmorphy equivalence classes of a given object $\rho(\mathbf{r})$ form the *symmorphy group* h_ρ of the object $\rho(\mathbf{r})$. This group h_ρ is formally defined [43,108] as the quotient group of group g_ρ with respect to the symmorphy equivalence $\boldsymbol{m}_\rho$. The product of two symmorphy classes T_1 and T_2 in the symmorphy group h_ρ is defined as the class T_3,

$$T_3 = T_1 \cdot T_2 \tag{8.22}$$

which contains all the products of form $t_1 t_2$ of any two transformations t_1 and t_2 from classes T_1 and T_2, respectively.

The symmorphy groups are generalizations of the point symmetry groups. This generalization is twofold: on the one hand, from discrete point sets to continua of general shapes that may show no symmetry properties at all and, on the other hand, from the algebraic structure of linear point symmetry operations to an algebraic structure of a much richer family of homeomorphisms of the 3D space, selected by the shape of the given object. The symmorphy group h_ρ of a continuum object such as the electronic charge density $\rho(\mathbf{r})$ provides a complete shape characterization of the object $\rho(\mathbf{r})$.

Most symmorphy groups h_ρ are rather complicated and their direct use for molecular shape characterization and shape similarity analysis is not a trivial task. Some simplifications are possible using a technique based on the Brouwer fixed point theorem, as described in reference [43].

Another simplification of more technical nature [108] is obtained using the FSGH method, outlined in Chapter 7. The electronic charge density function $\rho(\mathbf{r})$

is represented by a finite family of points, approximately uniformly distributed along a sequence of MIDCO's. The set of selected points along a MIDCO G(a) is denoted by P(a) and the entire 3D electronic density $\rho(\mathbf{r})$ is represented by the *collection* of point sets $P(a_i)$, i=1,2,...k, where each $P(a_i)$ set represents a $G(a_i)$ MIDCO for some selected a_i value. The actual density function $\rho(\mathbf{r})$ is replaced by the resulting collection of a finite number of points as the object defining the symmorphy transformations. If symmorphy equivalence and symmorphy groups are restricted to a finite point set, then the resulting algebraic structure is considerably simpler than that of the 3D continuum of the full charge density function $\rho(\mathbf{r})$. The fate of points of the space that are not included in the selection is immaterial during symmorphy transformations, since the selected points can transform only into one another. As it has been pointed out in [108], for such finite point sets the symmorphy group is isomorphic with a permutation group of those permutations as elements which interchange selected points that belong to the same MIDCO. This symmorphy group is a finite group that provides a full description of the shape of the selected point set, and a faithful description of the variation of the charge density from the $P(a_i)$ point set of one MIDCO $G(a_i)$ to that of another.

CLOSING REMARKS

The main goal of this book is to convince the reader that molecular shape is accessible to rigorous study, and that much can be learned from such analysis. However, molecular shape is just one special aspect of shape in chemistry. The concept of shape is universal and it is my hope that some of the ideas and methods presented in this book will find applications not only in molecular shape analysis but in a broader chemical context, perhaps beyond chemistry and perhaps beyond natural sciences. Within the sciences, shape is a powerful interdisciplinary subject, as witnessed by the rich variety of applications described in Forma, the official journal of The Society for Science on Form [451]. Some of the approaches originally developed for shape analysis can lead to unexpected interrelations between symmetry and energy [452], and to conjectures and open problems in mathematical chemistry [453]. Shape is useful, and shape is beautiful. Felix Klein, the mathematician of The Erlangen Program fame, was intrigued by the connections between shape and beauty: he drew all the lines of parabolic points (in our terminology, the boundaries of D_μ domains of ordinary convexity) on the surfaces of sculptures [454], and tried to find regularities in order to explain the esthetically pleasing aspects of shape. One of the sculptures he analyzed, the bust of the Apollo of Belvedere with lines of parabolic points marked, still can be found in the Institute of Mathematics of the Göttingen University. Klein did not publish his thoughts and apparently gave up the idea, but his study is referred to in a work by no lesser mathematicians than D. Hilbert and S. Cohn-Vossen [454]. Whether mathematics will ever be able to grasp the essence of beauty or not, the concept of shape is one of the most fundamental connections between the sciences and arts. Shape is everywhere.

REFERENCES

[1]   J.H. van't Hoff, *A Suggestion Looking to the Extension into Space of the Structural Formulas at Present Used in Chemistry - and a Note upon the Relation between the Optical Activity and the Chemical Constitution of Organic Compounds (in Dutch).* Greven, Utrecht, 1874.

[2]   J.H. van't Hoff, *The Arrangements of Atoms in Space,* 2nd ed., (transl. A. Eiloart), Longmans, Green, and Co., London, 1898.

[3]   J.A. Le Bel, *Bull. Soc. Chim. France,* **22**, 337 (1874).

[4]   K. Freudenberg, *Stereochemie.* F. Deuticke, Leipzig und Wien, 1932.

[5]   W. Klyne, *Progress in Stereochemistry.* Academic, New York, 1954.

[6]   M.S. Newman, *Steric Effects in Organic Chemistry.* John Wiley & Sons, New York, 1956.

[7]   E.L. Eliel, *Stereochemistry of Carbon Compounds.* McGraw-Hill, New York, 1962.

[8]   J. Grundy, *Stereochemistry: the Static Principles.* Butterworths, London, 1964.

[9]   K. Mislow, *Introduction to Stereochemistry.* Benjamin, New York, 1966.

[10]   N.L. Allinger and E.L. Eliel (Eds.), *Topics in Stereochemistry, Vol.1.* Wiley, New York, 1967.

[11]   E.L. Eliel, *Elements of Stereochemistry.* Wiley, New York, 1969.

[12]   D. Johannes, *Stereochemistry and Conformational Analysis.* Universitetsforlaget, Oslo, 1978.

[13]   O.B. Ramsay, *Stereochemistry.* Heydon, London, 1981.

[14]   M. Nógrádi, *Stereochemistry: Basic Concepts and Applications.* Pergamon Press, Oxford, 1981.

[15]   J. Rétey and J.A. Robinson, *Stereospecificity in Organic Chemistry and Enzymology.* Verlag Chemie, Weinheim, 1982.

[16]   E.L. Eliel and S.H. Wilen (Eds.), *Topics in Stereochemistry, Vol.17.* Wiley, New York, 1987.

[17]   J.H. van 't Hoff, *Bull. Soc. Chim. Fr.,* **23**, 295 (1875).

[18]   P. Curie, *J. Phys. (Paris),* **3**, 393 (1894).

[19]   R.S. Cahn, C.K. Ingold, and V. Prelog, *Angew. Chem., Internat. Ed. Engl.,* **5**, 385 (1966).

[20]   E. Ruch and A. Schönhofer, *Theor. Chim. Acta* **10**, 91 (1968).

[21]   E. Ruch, *Theor. Chim. Acta* **11**, 183 (1968).

[22]   V. Prelog and G. Helmchen, *Helv. Chim. Acta,* **55**, 2581 (1972).

[23]   A. Mead, E. Ruch, and A. Schönhofer, *Theor. Chim. Acta,* **29**, 269 (1973).

[24]   P. Decker, *J. Mol. Evol.,* **4**, 49 (1974).

[25]   J.G. Nourse and K. Mislow, *J. Amer. Chem. Soc.,* **97**, 4571 (1975).

[26]   R.W. Robinson, F. Harary, and A.T. Balaban, *Tetrahedron* **32**, 355 (1976).

[27]   K. Mislow and P. Bickart, *Isr. J. Chem.,* **15**, 1 (1976).

[28]   R.G. Woolley, *Adv. Phys.*, **25**, 27 (1976).

[29]   E. Ruch, *Angew. Chem., Int. Ed. Engl.*, **16**, 65 (1977).

[30]   D.J. Klein and A.H. Cowley, *J. Amer. Chem. Soc.,* **100**, 2593 (1978).

[31]   L.D. Barron, *J. Amer. Chem. Soc.*, **101**, 269 (1979).

[32]   J. Dugundji, J. Showell, R. Kopp, D. Marquarding, and I. Ugi, *Isr. J. Chem.,* **20**, 20 (1980).

[33]   R.L. Flurry, Jr., *J. Amer. Chem. Soc.,* **103**, 2901 (1981).

[34]   R.G. Woolley, *Struct. Bonding*, **52**, 1 (1982).

[35]   E. Ruch and D.J. Klein, *Theor. Chim. Acta* **63**, 447 (1983).

[36]   F.A.L. Anet, S.S. Miura, J. Siegel and K. Mislow, *J. Amer. Chem. Soc.,* **105**, 1419 (1983).

[37]   C.E. Wintner, *J. Chem. Educ.*, **60**, 550 (1983).

[38]   K. Mislow and J. Siegel, *J. Amer. Chem. Soc.,* **106**, 3319 (1984).

[39]   I. Ugi, J. Dugundji, R. Kopp and D. Marquarding, *Perspectives in Theoretical Stereochemistry (Lecture Notes in Chemistry, Vol. 36).* Springer-Verlag, New York, 1984, Chapters I, VI.

[40]   L.D. Barron, *Chem. Soc. Rev.*, **15**, 189 (1986).

[41]   W.A. Bonner, "Origins of Chiral Homogeneity in Nature", in *Topics in Stereochemistry, Vol. 18,* E.L. Eliel and S.H. Wilen (Eds.), Wiley, New York, 1988.

[42]   P. Busch and F.E. Schroeck, Jr., *Found. Phys.*, **19**, 807 (1989).

[43]   P.G. Mezey, "Topology of Molecular Shape and Chirality", in *New Theoretical Concepts for Understanding Organic Reactions,* J. Bertran and I.G. Csizmadia (Eds.), Nato ASI Series, Kluwer Academic Publishers, Dordrecht, 1989.

[44]   R.G. Woolley, *J. Mol. Struct. (Theochem)*, **232**, 13, (1991).

[45]   R.A. Hegstrom, *J. Mol. Struct. (Theochem)*, **232**, 17 (1991).

[46]   A.I. Kitaigorodskii, *Organic Chemical Crystallography,* Consultants Bureau, New York, 1961, Chap. 4.

[47]   A. Rassat, *Compt. rend. Acad. Sci. (Paris)*, II **299**, 53 (1984).

[48]   G. Gilat, *Chem. Phys. Lett.*, **121**, 9 (1985).

[49]   G. Gilat and L.S. Schulman, *Chem. Phys. Lett.*, **121**, 13 (1985).

[50]   G. Gilat, *J. Phys. A*, **22**, L545 (1989).

[51]   A.B. Buda, T.P.E. Auf der Heyde, and K. Mislow, *J. Math. Chem.*, **6**, 243 (1991).

[52]   T.P.E. Auf der Heyde, A.B. Buda, and K. Mislow, *J. Math. Chem.*, **6**, 255 (1991).

[53]   A.B. Buda and K. Mislow, *Elem. Math.*, **46**, 65 (1991).

[54]   F. Harary and P.G. Mezey, "Chiral and Achiral Square-Cell Configurations and the Degree of Chirality", in *New Developments in Molecular Chirality*, P.G. Mezey (Ed.), Kluwer Academic Publishers, Dordrecht, 1991.

[55]   P.G. Mezey, "A Global Approach to Molecular Chirality", in *New Developments in Molecular Chirality,* P.G. Mezey (Ed.), Kluwer Academic Publishers, Dordrecht, 1991.

[56] D. Avnir and A.Y. Meyer, *J. Mol. Struct. (Theochem)*, **226**, 211 (1991).

[57] A.B. Buda and K. Mislow, *J. Mol. Struct. (Theochem)*, **232**, 1 (1991).

[58] A.B. Buda, T. Auf der Heyde, and K. Mislow, *Angew. Chem., Int. Ed. Engl.*, **31**, 989 (1992).

[59] D.M. Walba, "Stereochemical Topology", in *Chemical Applications of Topology and Graph Theory*, R.B. King (Ed.), Elsevier, Amsterdam, 1983.

[60] D.M. Walba, Tetrahedron, **41**, 3161 (1985).

[61] K.C. Millett, *Croat. Chem. Acta*, **59**, 669 (1986).

[62] P.G. Mezey, *J. Amer. Chem. Soc.*, **108**, 3976 (1986).

[63] D.W. Sumners, *J. Math. Chem.*, **1**, 1 (1987).

[64] J. Simon, *Topology*, **25**, 229 (1986).

[65] E. Flapan, *Pac. J. Math.*, **129**, 57 (1987).

[66] D.M. Walba in *Graph Theory and Topology in Chemistry*, R.B. King and D.H. Rouvray (Eds.), Elsevier, Amsterdam, 1987, p.23.

[67] J. Simon in *Graph Theory and Topology in Chemistry*, R.B. King and D.H. Rouvray (Eds.), Elsevier, Amsterdam, 1987.

[68] J. Simon, *J. Comput. Chem.*, **8**, 718 (1987).

[69] E. Flapan in *Graph Theory and Topology in Chemistry*, R.B. King and D.H. Rouvray (Eds.), Elsevier, Amsterdam, 1987, p.76.

[70] K.C. Millett in *New Developments in Molecular Chirality*, P.G. Mezey (Ed.), Kluwer Academic Publishers, Dordrecht, 1991, p. 165.

[71] E. Flapan in *New Developments in Molecular Chirality*, P.G. Mezey (Ed.), Kluwer Academic Publishers, Dordrecht, 1991, p. 209.

[72] D.M. Walba in *New Developments in Molecular Chirality*, P.G. Mezey (Ed.), Kluwer Academic Publishers, Dordrecht, 1991, p.119.

[73] J.D. Donaldson and S.D. Ross, *Symmetry and Stereochemistry*. Wiley, New York, 1972.

[74] S.F.A. Kettle, *Symmetry and Structure*. Wiley, New York, 1985.

[75] I. Hargittai and M. Hargittai, *Symmetry Through the Eyes of a Chemist*. VCH Publishers, New York, 1986.

[76] I. Hargittai (Ed.), *Symmetry 2: Unifying Human Understanding*. Pergamon Press, New York, 1989.

[77] I. Hargittai (Ed.), *Crystal Symmetries*. Pergamon Press, Oxford, 1989.

[78] I. Hargittai (Ed.), *Quasicrystals, Networks, and Molecules of Fivefold Symmetry*. VCH Publishers, New York, 1990.

[79] F.A. Cotton, *Chemical Applications of Group Theory*, 3rd ed. Wiley-Interscience, New York, 1990.

[80] J.E. Leonard, G.S. Hammond, and H.E. Simmons, *J. Amer. Chem. Soc.*, **97**, 5052 (1975).

[81] P. Murray-Rust, H.-B. Bürgi, and J.D. Dunitz, *Acta Cryst.*, B**34**, 1787, (1978).

[82] P. Murray-Rust, H.-B. Bürgi, and J.D. Dunitz, *Acta Cryst.*, A**35**, 703 (1979).

[83] E.L. Eliel, N.L. Allinger, S.J. Angyal, and G.A. Morrison, *Conformation Analysis*. Wiley, New York, 1965.

[84] H.A. Stuart, *Z. phys. Chem.*, B 27, 350 (1927).

[85]   F. M. Richards, *Annu. Rev. Biophys. Bioeng.,* **6**, 151 (1977).

[86]   A. Gavezotti, *J. Amer. Chem. Soc.,* **105**, 5220 (1983).

[87]   M.L. Connolly, *J. Amer. Chem. Soc.,* **107**, 1118 (1985).

[88]   K. D. Gibson and H. A. Scheraga, *Mol. Phys.,* **62**, 1247 (1987).

[89]   P. Coppens and M. B. Hall, Eds., *Electron Distribution and the Chemical Bond,* Plenum, New York and London, 1982.

[90]   S. Fliszár, *Charge Distributions and Chemical Effects,* Springer, New York, 1983.

[91]   G.D. Purvis III and C. Culberson, *Int. J. Quantum Chem., Quantum Biol. Symp.,* **13**, 261 (1986).

[92]   P.A. Kollman, *J. Amer. Chem. Soc.,* **100**, 2974 (1978).

[93]   R.J. Gillespie, *Molecular Geometry.* Van Nostrand Reinhold, London, 1972.

[94]   R.J. Gillespie and I. Hargittai, *The VSEPR Model of Molecular Geometry.* Allyn and Bacon, Boston, 1991.

[95]   R.B. Woodward and R. Hoffmann, *Angew. Chem. Int. Ed. Engl.,* **8**, 781 (1969).

[96]   K. Fukui, T. Yonezawa, and H. Shingu, *J. Chem. Phys.,* **20**, 722 (1952).

[97]   K. Fukui, in *Molecular Orbitals in Chemistry, Physics, and Biology.* P.-O. Löwdin and B. Pullman (Eds.), Academic Press, New York, 1964.

[98]   J.K. Burdett, *Molecular Shapes.* Wiley, New York, 1980.

[99]   Z. Simon, A. Chiriac, S. Holban, D. Ciubotaru, and G.I. Mihalas, *Minimum Steric Difference. The MTD Method for QSAR Studies.* Research Studies Press, Ltd., Letchworth, Hertfordshire, England, and Wiley, New York, 1984.

[100] R. Hoffmann and P. Laszlo, *Angew. Chem. Int. Ed. Engl.,* **30**, 1 (1991).

[101] R.G. Woolley, *J. Amer. Chem. Soc.,* **100**, 1073 (1978).

[102] R.G. Woolley, *New Scientist,* 53 (Oct. 22, 1988).

[103] P.G. Mezey, "Topological Theory of Molecular Conformations," in *Structure and Dynamics of Molecular Systems,* R. Daudel, J.-P. Korb, J.-P. Lemaistre, and J. Maruani (Eds.), Reidel, Dordrecht, 1985.

[104] P.G. Mezey, "Topological Model of Reaction Mechanisms," in *Structure and Dynamics of Molecular Systems,* R. Daudel, J.-P. Korb, J.-P. Lemaistre, and J. Maruani (Eds.), Reidel, Dordrecht, 1985.

[105] P.G. Mezey, "Reaction Topology," in *Applied Quantum Chemistry,* Proceedings of the Hawaii 1985 Nobel Laureate Symposium on Applied Quantum Chemistry, V.H. Smith, Jr., H.F. Schaefer III, and K. Morokuma (Eds.), Reidel, Dordrecht, 1986, pp. 53-74.

[106] P.G. Mezey, *Potential Energy Hypersurfaces.* Elsevier, Amsterdam, 1987.

[107] P.G. Mezey, "Topological Quantum Chemistry," in *Reports in Molecular Theory,* G. Náray-Szabó and H. Weinstein (Eds.), CRC Press, Boca Raton, 1990.

[108] P.G. Mezey, "Three-Dimensional Topological Aspects of Molecular Similarity," in *Concepts and Applications of Molecular Similarity,* M. A. Johnson and G.M. Maggiora (Eds.), Wiley, New York, 1990.

[109] P.G. Mezey, "Molecular Surfaces," in *Reviews in Computational Chemistry,* K.B. Lipkowitz and D.B. Boyd (Eds.), VCH Publishers, New York, 1990.

[110] P.G. Mezey, "The Topology of Molecular Surfaces and Shape Graphs", in *Computational Chemical Graph Theory"*, D.H. Rouvray (Ed.), Nova, New York, 1990.

[111] P.G. Mezey, "Non-Visual Molecular Shape Analysis: Shape Changes in Electronic Excitations and Chemical Reactions," in *Computational Advances in Organic Chemistry (Molecular Structure and Reactivity)*, C. Ogretir and I. Csizmadia (Eds.), Kluwer Academic Publishers, Dordrecht, 1991.

[112] G.A. Arteca and P.G. Mezey, "Algebraic Approaches to the Shape Analysis of Biological Macromolecules," in *Computational Chemistry: Structure, Interactions, and Reactivity, Part A,* S. Fraga (Ed.), Elsevier, Amsterdam, 1992.

[113] G.F. Simmons, *Introduction to Topology and Modern Analysis.* McGraw-Hill, New York, 1963.

[114] T.W. Gamelin and R.E. Greene, *Introduction to Topology.* Saunders College Publishing, New York, 1963.

[115] I.M. Singer and J.A. Thorpe, *Lecture Notes on Elementary Topology and Geometry.* Springer-Verlag, New York, 1976.

[116] M. Morse and S.S. Cairns, *Critical Point Theory in Global Analysis and Differential Topology: an Introduction.* Academic Press, New York, London, 1969.

[117] V. Guillemin and A. Pollack, *Differential Topology.* Prentice Hall, Englewood Cliffs, 1974.

[118] R.L. Bishop and R.J. Crittenden, *Geometry of Manifolds.* Academic Press, New York, 1964.

[119] E.H. Spanier, *Algebraic Topology.* McGraw-Hill, New York, 1966.

[120] M. Greenberg, *Lectures on Algebraic Topology.* Benjamin, New York, 1967.

[121] S.-T. Hu, *Elements of General Topology.* Holden-Day, San Francisco, 1969.

[122] J. Vick, *Homology Theory.* Academic Press, New York, 1973.

[123] K. Borsuk, *Fund. Math.,* **62**, 223 (1968).

[124] K. Borsuk, *Fund. Math.,* **67**, 221 (1970).

[125] K. Borsuk, *Fund. Math.,* **67**, 265 (1970).

[126] K. Borsuk, *Colloq. Math.,* **21**, 247 (1970).

[127] K. Borsuk, *Bull. Acad. Polon. Sci. Ser. Sci. Math. Astronom. Phys.,* **18**, 127 (1970).

[128] W. Holsztynski, *Fund. Math.,* **70**, 157 (1971).

[129] T.A. Chapman, *Fund. Math.,* **76**, 181 (1972).

[130] T.J. Sanders, *Pacific J. Math.,* **48**, 485 (1973).

[131] K. Borsuk, *Fund. Math.,* **85**, 185 (1974).

[132] K. Borsuk, *Theory of Shape. Monografie Matematyczne,* Vol. **59**. Polish Scientific Publishers, Warsawa, 1975.

[133] P. Bacon, *Proc. Amer. Math. Soc.,* **53**, 489 (1975).

[134] L.C. Siebenmann, *Manuscripta Math.,* **16**, 373 (1975).

[135] D.A. Edwards and R. Geoghegan, *Trans. Amer. Math. Soc.,* **222**, 389 (1976).

[136] J.E. Keesling, *Fund. Math.*, **92**, 195 (1976).

[137] T.J. Sanders, *Fund. Math.*, **93**, 37 (1976).

[138] J. Segal, *Shape Theory Notes. Mimeographed notes.* University of Washington, 1976.

[139] D.A. Edwards and P.T. McAuley, *Fund. Math.*, **96**, 195 (1977).

[140] Vo-Thanh-Liem, *Fund. Math.*, **97**, 221 (1977).

[141] F. Wattenberg, *Fund. Math.*, **98**, 41 (1978).

[142] S. Mardesic, *The Foundations of Shape Theory. Lecture notes.* University of Kentucky, 1978.

[143] M. Moszynska, *Fund. Math.*, **99**, 15 (1978).

[144] J. Dydak and J. Segal, *Shape Theory: An Introduction. Lecture Notes in Mathematics* **688**. Springer-Verlag, Berlin, 1978.

[145] J.L. MacDonald, *J. Pure Appl. Algebra*, **12**, 79 (1978).

[146] R.R. Summerhill, *A Categorical Description of Shape Theory. Mimeographed Notes.* Kansas State University, Manhattan, 1979.

[147] T. Watanabe, *Fund. Math.*, **104**, 1 (1979).

[148] Y. Kodama and J. Ono, *Fund. Math.*, **105**, 29 (1979).

[149] Y. Kodama and J. Ono, *Fund. Math.*, **108**, 29 (1980).

[150] A. Calder and J. Siegel, *J. Pure Appl. Algebra*, **20**, 129 (1981).

[151] S. Zdravkovska, *Proc. Amer. Math. Soc.*, **83**, 594 (1981).

[152] S. Ungar, *Topology Appl.*, **12**, 89 (1981).

[153] S. Mardesic and J. Segal, *Shape Theory. (The Inverse System Approach).* North-Holland Publishers, Amsterdam, 1982.

[154] S. Mardesic and J. Segal (Eds.), *Geometric Topology and Shape Theory. Lecture Notes in Mathematics* Vol. **1283**. Springer-Verlag, Berlin, 1987.

[155] P.G. Mezey, *Int. J. Quantum Chem. Quant. Biol. Symp.*, **12**, 113 (1986).

[156] P.G. Mezey, *J. Comput. Chem.*, **8**, 462 (1987).

[157] P.G. Mezey, *Int. J. Quantum Chem. Quant. Biol. Symp.*, **14**, 127 (1987).

[158] P.G. Mezey, *J. Math. Chem.*, **2**, 299 (1988).

[159] F. Harary, *Graph Theory.* Addison-Wesley, Reading, Mass., 1969.

[160] Y.C. Martin, *Quantitative Drug Design: A Critical Introduction.* Dekker, New York, 1978.

[161] M. Le Bret, *Biopolymers* , **18**, 1709 (1979).

[162] C.R. Cantor and P.R. Schimmel, *The Conformation of Biological Macromolecules, Biophysical Chemistry, Part I.* Freeman, San Francisco, 1980.

[163] A.M. Lesk and K.D. Hardman, *Science*, **216**, 539 (1982).

[164] M. Karplus and J.A. McCammon, *Annu. Rev. Biochem.*, **53**, 263 (1983).

[165] P. De Santis, S. Morosetti, and A. Palleschi, *Biopolymers*, **22**, 37 (1983).

[166] W.G. Richards, *Quantum Pharmacology.* Butterworths, London, 1983.

[167] R. Franke, *Theoretical Drug Design Methods.* Elsevier, Amsterdam, 1984.

[168] G.E. Schulz and R.H. Schirmer, *Principles of Protein Structure.* Springer-Verlag, New York, 1979.

[169] J.S. Richardson, *Methods in Enzymol.*, **115**, 359 (1985).

[170] M.N. Liebman, C.A. Venanzi, and H. Weinstein, *Biopolymers*, **24**, 1721 (1985).

[171] C.J. Rawlings, W.R. Taylor, J. Nyakairu, J. Fox, and M.J.E. Sternberg, *J. Mol. Graph.*, **3**, 151 (1985).

[172] A.M. Lesk and K.D. Hardman, *Methods in Enzymol.*, **115**, 381 (1985).

[173] T. Kikuchi, G. Némethy, and H.A. Scheraga, *J. Comput. Chem.*, **7**, 67 (1986).

[174] M. Carson and C.E. Bugg, *J. Mol. Graph.*, **4**, 121 (1986).

[175] P.M. Dean, *Molecular Foundations of Drug-Receptor Interaction.* Cambridge University Press, New York, 1987.

[176] M. Carson, *J. Mol. Graph.*, **5**, 103 (1987).

[177] M.L. Connolly, *Visual Comput.*, **3**, 72 (1987).

[178] R. Jaenicke, *Prog. Biophys. Molec. Biol.*, **49**, 117 (1987).

[179] O. Tapia, H. Eklund, and C.I. Brändén, "Molecular, Electronic, and Structural Aspects of the Catalytic Mechanism of Alcohol Dehydrogenase," in *Steric Aspects of Biomolecular Interactions*, G. Náray-Szabó and K. Simon (Eds.), CRC Press, West Palm Beach, 1987.

[180] J. Åqvist and O. Tapia, *J. Mol. Graph.*, **5**, 30 (1987).

[181] S.E. Leicester, J.L. Finney, and R.P. Bywater, *J. Mol. Graph.*, **6**, 104 (1988).

[182] F.M. Richards and C.E. Kundot, *Protein Struct. Funct. Genet.*, **3**, 71 (1988).

[183] R.A. Abagyan, and V.N. Maiorov, *J. Biomol. Struct. Dynam.*, **5**, 1267 (1988).

[184] T. Dearden, *J. Comput. Chem.*, **10**, 529 (1989).

[185] M.-H. Hao and W.K. Olson, *Biopolymers,* **28**, 873 (1989).

[186] E.M. Mitchell, P.J. Artymiuk, D.W. Rice, and P. Willett, *J. Mol. Biol.,* **212**, 151 (1990).

[187] C.L. Fisher, J.A. Tainer, M.E. Pique, and E.D. Getzoff, *J. Mol. Graph.*, **8**, 125 (1990).

[188] N. Colloc'h and J.P. Mornon, *J. Mol. Graph.*, **8**, 13. (1990).

[189] D. Wang, H.P.C. Driessen, and I.J. Tickle, *J. Mol. Graph.*, **9**, 50 (1991).

[190] G.A. Arteca and P.G. Mezey, *Int. J. Quantum Chem. Quant. Biol. Symp.*, **14**, 133 (1987).

[191] G.A. Arteca, V.B. Jammal, P.G. Mezey, J.S. Yadav, M.A. Hermsmeier, and T.M. Gund, *J. Mol. Graph.*, **6**, 45 (1988).

[192] G.A. Arteca, V.B. Jammal, and P.G. Mezey, *J. Comput. Chem.*, **9**, 608 (1988).

[193] P.G. Mezey, *IEEE Eng. Med. Bio. Soc. 11th Annual Int. Conf.*, **11**, 1905 (1989).

[194] G.A. Arteca and P.G. Mezey, *Int. J. Quantum Chem.*, **34**, 517 (1988).

[195] G.A. Arteca and P.G. Mezey, *J. Comput. Chem.*, **9**, 554 (1988).

[196] P.G. Mezey, "Dynamic Shape Analysis of Biomolecules Using Topological Shape Codes", in *The Role of Computational Models and Theories in Biotechnology.* J. Bertran (Ed.), pp. 83-104, Kluwer Academic Publishers, Dordrecht, 1992.

[197] G.A. Arteca and P.G. Mezey, *J. Mol. Graphics,* **8**, 66 (1990).

[198] G.A. Arteca, O. Tapia, and P.G. Mezey, *J. Mol. Graphics,* **9**, 148 (1991).

[199] P.G. Mezey, *J. Math. Chem.,* **2**, 325 (1988).

[200] J.J. Berzelius, *Jahresber. Tübingen,* **20**, 13 (1840).

[201] A.G. Murzin and A.V. Finkelstein, *J. Mol. Biol.,* **204**, 749 (1988).

[202] C. Chothia, *Nature,* **337**, 204 (1989).

[203] G.M. Maggiora, P.G. Mezey, B. Mao, and K.C. Chou, *Biopolymers,* **30**, 211 (1990).

[204] H.L. Frisch and E. Wasserman, *J. Amer. Chem. Soc.,* **83**, 3789 (1961).

[205] W. Vetter and G. Schill, *Tetrahedron,* **23**, 3079 (1967).

[206] G. Schill, "Catenanes, Rotaxanes and Knots", in *Organic Chemistry, A Series of Monographs,* Vol. 22, A.T. Blomquist (Ed.), Academic Press, New York, London, 1971.

[207] H. Fritz, P. Hug, H. Sauter, and E. Logemann, *J. Magn. Reson.*, **21**, 373 (1976).

[208] G. Schill, E. Logemann, and W. Littke, *Chemie in uns. Zeit,* **18**, 130 (1984).

[209] S.J. Spengler, A. Stasiak, and N.R. Cozzarelli, *Cell,* **42**, 325 (1985).

[210] F.B. Dean, A. Stasiak, T. Koller, and N.R. Cozzarelli, *J. Biol. Chem.*, **260**, 4795 (1985).

[211] M. Cesario, C.O. Dietrich-Buchecker, J. Guilhem, C. Pascard, and J.-P. Sauvage, *J. Chem. Soc. Chem. Commun.,* 244 (1985).

[212] J. Roovers, *Macromolecules,* **18**, 1359 (1985).

[213] S.A. Wasserman and N.R. Cozzarelli, *Science,* **232**, 951 (1986).

[214] J.-M. Lehn and C. Sirlin, *New J. Chem.,* **11**, 693 (1987).

[215] J.-M. Lehn and P. G. Potvin, *Can. J. Chem.,* **66**, 195 (1988).

[216] J.-M. Lehn, *Angew. Chem.,* **100**, 91 (1988).

[217] J.-M. Lehn, M. Mascal, A. DeCian, and J. Fischer, *J. Chem. Soc., Chem. Commun.,* **1990**, 479 (1990).

[218] K. Schoetz, T. Clark, and P.v.R. Schleyer, *J. Amer. Chem. Soc.,* **110**, 1394 (1988).

[219] C. Schade, P.v.R. Schleyer, P. Gregory, H. Dietrich, and W. Mahdi, *J. Organomet. Chem.,* **341**, 19 (1988).

[220] J.W. Bausch, P.S. Gregory, G.A. Oláh, G.K.S. Prakash, P.v.R. Schleyer, and G.A. Segal, *Amer. Chem. Soc.,* **111**, 3633 (1989).

[221] C.O. Dietrich-Buchecker, J. Guilhem, A.K. Khemiss, J.P. Kintzinger, C. Pascard, and J.-P. Sauvage, *Angew. Chem. Int. Ed.*, **26**, 661 (1987).

[222] C.O. Dietrich-Buchecker and J.-P. Sauvage, *Chem. Rev.*, **87**, 795 (1987).

[223] D.K. Mitchell and J.-P. Sauvage, *Angew. Chem. Int. Ed. Engl.*, **27**, 930 (1988).

[224] G. Schill, N. Schweickert, H. Fritz, and W. Vetter, *Chem. Ber.*, **121**, 961 (1988).

[225] C.O. Dietrich-Buchecker and J.-P. Sauvage, *Angew. Chem. Int. Ed. Engl.*, **28**, 189 (1989).

[226] P.R. Ashton, T.T. Goodnow, A.E. Kaifer, M.V. Reddington, A.M.Z. Slawin, N. Spencer, J.F. Stoddart, C. Vicent, and D.J. Williams, *Angew. Chem. Int. Ed. Engl.*, **28**, 1396 (1989).

[227] C.O. Dietrich-Buchecker, J. Guilhem, C. Pascard, and J.-P. Sauvage, *Angew. Chem. Int. Ed. Engl.*, **29**, 1154 (1990).

[228] J.-P Sauvage, *Acc. Chem. Res.,* **23**, 319 (1990).

[229] C. Seel and F. Vögtle, *Angew. Chem. Int. Ed. Engl.,* **31**, 528 (1992).

[230] P. Lucio Anelli, P.R. Ashton, R. Ballardini, V. Balzani, M. Delgado, M.R. Gandolfi, T.T. Goodnow, A.E. Kaifer, D. Philp, M. Pietraszkiewicz, L. Prodi, M.V. Reddington, A.M.Z. Slawin, N. Spencer, J.F. Stoddart, C. Vicent, and D.J. Williams, *J. Amer. Chem. Soc.,* **114**, 193 (1992).

[231] R.P. Sijbesma and R.J.M. Nolte, *J. Amer. Chem. Soc.,* **113**, 6695 (1991).

[232] A. Bencini, A. Bianchi, M.I. Burguete, E. García-España, S.V. Luis, and J.A. Ramírez, *J. Amer. Chem. Soc.,* **114**, 1919 (1992).

[233] L. Pang and M.A. Whitehead, *Supramol. Chem.,* **1**, 81 (1992).

[234] T. Tjivikua, P. Ballester, and J. Rebek, Jr., *J. Amer. Chem. Soc.,* **112**, 1249 (1990).

[235] V. Rotello, J.-I. Hong, and J. Rebek, Jr., *J. Amer. Chem. Soc.,* **113**, 9422 (1991).

[236] J.S. Nowick, Q. Feng, T. Tjivikua, P. Ballester, and J. Rebek, Jr., *J. Amer. Chem. Soc.,* **113**, 8315 (1991).

[237] J. Rebek, Jr., *Science,* **256**, 1179 (1992).

[238] T.K. Park, Q. Feng, and J. Rebek, Jr., *J. Amer. Chem. Soc.,* **114**, 4529 (1992).

[239] A. Terfort and G. von Kiedrowski, *Angew. Chem. Int. Ed. Engl.,* **31**, 654 (1992).

[240] P.G. Mezey, *J. Math. Chem.,* **11**, 27 (1992).

[241] F. Hausdorff, *Set Theory*; transl. by J.R. Auman *et al.* Chelsey, New York, 1957, pp. 166-168.

[242] A.B. Buda and K. Mislow, *J. Amer. Chem. Soc.,* **114**, 6006 (1992).

[243] P.G. Mezey, *J. Math. Chem.,* **7**, 39 (1991).

[244] V.F.R. Jones, *Bull. Amer. Math. Soc. (NS),* **12**, 103 (1985).

[245] J.A. Pople, *J. Amer. Chem. Soc.,* **102**, 4615 (1980).

[246] P. Pechukas, *J. Chem. Phys.,* **64**, 1516 (1976).

[247] P.G. Mezey, *Can. J. Chem.,* **70**, 343 (1992).

[248] P.G. Mezey, *J. Math. Chem.,* **13**, 59 (1993).

[249] P.G. Mezey, *Int. J. Quantum Chem.,* **25**, 853 (1984).

[250] M. Born and R. Oppenheimer, *Ann. Phys.,* **84**, 457 (1927).

[251] P.G. Mezey, "From Geometrical Molecules to Topological Molecules: A Quantum Mechanical View", in *Molecules in Physics, Chemistry and Biology,* Vol II, J. Maruani (Ed.), pp. 61-81, Reidel, Dordrecht, 1988.

[252] P.G. Mezey and J. Maruani, *Mol. Phys.,* **69**, 97 (1990).

[253] M.J. Frisch, M. Head-Gordon, G.W. Trucks, J.B. Foresman, H.B. Schlegel, K. Raghavachari, M.A. Robb, J.S. Binkley, C. González, D.J. Defrees, D.J. Fox, R.A. Whiteside, R. Seeger, C.F. Melius, J. Baker, R.L. Martin, L.R. Kahn, J.J.P. Stewart, S. Topiol and J.A. Pople, GAUSSIAN 90; Carnegie-Mellon Quantum Chemistry Publishing Unit, Pittsburgh, PA, 1990.

[254] P.D. Walker, G.A. Arteca, and P.G. Mezey, GSHAPE 90; Mathematical Chemistry Research Unit, University of Saskatchewan, Saskatoon, Canada, 1990.

[255] G.A. Arteca, N.D. Grant, and P.G. Mezey, *J. Comput. Chem.,* **12**, 1198 (1991).

[256] S.F. Boys, in *Quantum Theory of Atoms, Molecules and the Solid State*, P.-O. Löwdin (Ed.), Academic Press, New York, 1966.

[257] C. Edmiston and K. Ruedenberg, in *Quantum Theory of Atoms, Molecules and the Solid State*, P.-O. Löwdin (Ed.), Academic Press, New York, 1966.

[258] R. Kari, *Int. J. Quantum Chem.*, **25**, 321 (1984).

[259] J. Pipek and P.G. Mezey, *Int. J. Quantum Chem. Symp.*, **22**, 1 (1988).

[260] J. Pipek and P.G. Mezey, *J. Chem. Phys.*, **90**, 4916 (1989).

[261] Th. Berlin, *J. Chem. Phys.*, **19**, 208 (1951).

[262] P.G. Mezey, *J. Chem. Inf. Comp. Sci*, **32**, 650 (1992).

[263] P.D. Walker, G. Heal, M. Ramek, L. Raju, K. Gilbertson, and P.G. Mezey (unpublished results).

[264] J.-E. Dubois and P.G. Mezey, *Int. J. Quantum Chem.*, **43**, 647 (1992).

[265] K. Collard and G.G. Hall, *Int. J. Quantum Chem.*, **12**, 623 (1977).

[266] Y. Tal, R.F.W. Bader, T.T. Nguyen-Dang, M. Ojha, and S.G. Anderson, *J. Chem. Phys.*, **74**, 5162 (1981).

[267] R.F.W. Bader, T.S. Slee, D. Cremer, and E. Kraka, *J. Amer. Chem. Soc.*, **105**, 5061 (1983).

[268] D. Cremer and E. Kraka, *Croat. Chem. Acta*, **57**, 1265 (1984).

[269] R.F.W. Bader, *Acc. Chem. Res.*, **18**, 9 (1985).

[270] R.F.W. Bader, *Atoms in Molecules: A Quantum Theory*. Clarendon Press, Oxford, 1990.

[271] P.G. Mezey, *J. Chem. Phys.*, **78**, 6182 (1983).

[272] J. Cioslowski, *J. Phys. Chem.*, **94**, 5496 (1990).

[273] J. Cioslowski, S.T. Mixon, and W.D. Edwards, *J. Amer. Chem. Soc.*, **113**, 1083 (1991).

[274] J. Cioslowski and E.D. Fleischmann, *J. Chem. Phys.*, **94**, 3730 (1991).

[275] J. Cioslowski, P.B. O'Connor, and E.D. Fleischmann, *J. Amer. Chem. Soc.*, **113**, 1086 (1991).

[276] J. Cioslowski, S.T. Mixon, and E.D. Fleischmann, *J. Amer. Chem. Soc.*, **113**, 4751 (1991).

[277] J. Cioslowski and S.T. Mixon, *Can. J. Chem.*, **70**, 443 (1992).

[278] M. O'Keeffe and N.E. Brese, *J. Amer. Chem. Soc.*, **113**, 3226 (1991).

[279] G.G. Hoffman, R.A. Harris, and L.R. Pratt, *Can. J. Chem.*, **70**, 478, (1992).

[280] T. Koga, K. Sano, and T. Morita, *Theor Chim Acta*, **81**, 21 (1991).

[281] K.B. Wiberg, C.M. Hadad, T.J. LePage, C.M. Breneman, and M.J. Frisch, *J. Phys. Chem.*, **96**, 671 (1992).

[282] R. Rein, J.R. Rabinowitz, and T.J. Swissler, *J. Theor. Biol.*, **34**, 215 (1972).

[283] E. Scrocco and J. Tomasi, *Topics Current Chem.*, **42**, 95 (1973).

[284] D.M. Hayes and P.A. Kollman, *J. Amer. Chem. Soc.*, **98**, 3335 (1976).

[285] P.A. Kollman, *Acc. Chem. Res.*, **10**, 36 (1977).

[286] E. Scrocco and J. Tomasi, *Adv. Quantum Chem.*, **11**, 115 (1978).

[287] J. Tomasi, "On the Use of Electrostatic Molecular Potentials in Theoretical Investigations on Chemical Reactivity", in *Quantum Theory of Chemical Reactions*, Vol.1, R. Daudel, A. Pullman, L. Salem, and A. Veillard, (Eds.), Reidel, Dordrecht, 1979, pp.191-228.

[288] G. Náray-Szabó, *Quantum Chem. Program Exchange*, **13**, 396 (1980).

[289] P. Politzer and D.G. Truhlar (Eds.), *Chemical Applications of Atomic and Molecular Electrostatic Potentials*. Plenum, New York, 1981.

[290] G. Náray-Szabó, A. Grofcsik, K. Kósa, M. Kubinyi, and A. Martin, *J. Comput. Chem.*, **2**, 58 (1981).

[291] H. Weinstein, R. Osman, J.P. Green, and S. Topiol, "Electrostatic Potentials as Descriptors of Molecular Reactivity", in *Chemical Applications of Atomic and Molecular Electrostatic Potentials*, P. Politzer and D.G.Truhlar (Eds.), Plenum, New York, 1981, pp. 309-334.

[292] J.J. Kaufman, P.C. Hariharan, F.L. Tobin, and C. Petrongolo, "Electrostatic Molecular Potential Contour Maps from *Ab Initio* Calculation", in *Chemical Applications of Atomic and Molecular Electrostatic Potentials*, P. Politzer and D.G. Truhlar (Eds.), Plenum, New York, 1981, pp. 335-380.

[293] A. Pullman and B. Pullman, "Electrostatic Molecular Potential of the Nucleic Acids", in *Chemical Applications of Atomic and Molecular Electrostatic Potentials*, P. Politzer and D.G. Truhlar (Eds.), Plenum, New York, 1981.

[294] A. Warshel, *Acc. Chem. Res.*, **14**, 284 (1981).

[295] P.H. Reggio, H. Weinstein, R. Osman, and S. Topiol, *Int. J. Quantum Chem., Quantum Biol. Symp.* **8**, 373 (1981).

[296] S. Cox and D. Williams, *J. Comput. Chem.*, **2**, 304 (1981).

[297] Z. Gabányi, P. Surján, and G. Náray-Szabó, *Eur. J. Med Chem.*, **17**, 307 (1982).

[298] H. Weinstein, R. Osman, S. Topiol, and C.A. Venanzi, *Pharmacochem. Libr.*, **6**, 81 (1983).

[299] J. Angyán and G. Náray-Szabó, *J. Theor. Biol.*, **103**, 777 (1983).

[300] G. Náray-Szabó and P.R. Surján, "Computational Methods for Biological Systems", in *Theoretical Chemistry of Biological Systems, in Studies in Physical and Theoretical Chemistry*, Vol. 41, G. Náray-Szabó (Ed.), Elsevier, Amsterdam, 1986, pp.1-100.

[301] J.R. Rabinowitz and S.B. Little, *Int. J. Quantum Chem., Quantum Biol. Symp.*, **13**, 9 (1986).

[302] J.C. Culberson, G.D. Purvis III, M.C. Zerner, and B.A. Seiders, *Int. J. Quantum Chem., Quantum Biol. Symp.*, **13**, 267 (1986) .

[303] P.K. Weiner, R. Langridge, J.M. Blaney, R. Schaefer, and P.A. Kollman, *Proc. Natl. Acad. Sci. USA*, **79**, 3754 (1982).

[304] T.P. Lybrand, P.A. Kollman, V.C. Yu, and W. Sadee, *Pharm. Res.*, **3**, 218 (1986).

[305] J.R. Rabinowitz, K. Namboodiri, and H. Weinstein, *Int. J. Quantum Chem.*, **29**, 1697 (1986).

[306] W. Ito, H. Nakamura, and Y. Arata, *J. Mol. Graphics*, **7**, 60 (1989).

[307] D.S. Goodsell, I.S. Mian, and A.J. Olson, *J. Mol. Graphics*, **7**, 41 (1989).

[308] P. Sjoberg, J.S. Murray, T. Brinck, P. Evans, and P. Politzer, *J. Mol. Graphics*, **8**, 81 (1990).

[309] R.G. Parr and A. Berk, "The Bare-Nuclear Potential as Harbinger for the Electron Density in a Molecule", in *Chemical Applications of Atomic and Molecular Electrostatic Potentials*, P. Politzer and D.G.Truhlar (Eds.), Plenum, New York, 1981, pp. 51-62.

[310] B.Lee and F.M. Richards, *J. Mol. Biol.*, **55**, 379 (1971).
[311] R.P. Sheridan, R. Nilakantan, J.S. Dixon, and R. Venkataraghavan, *J. Med. Chem.*, **29**, 899 (1986).
[312] G.M. Crippen, *Distance Geometry and Conformational Calculations.* Research Studies Press, Chichester, UK, 1981.
[313] G.M. Crippen and T.F. Havel, *Distance Geometry and Molecular Conformation.* Wiley, New York, 1988.
[314] SYBYL, MENDYL; Tripos Associates Inc., St. Louis, MO 63117, USA.
[315] CHEM-X; Chemical Design Ltd., Oxford, U.K.
[316] FRODO; Rice University, Houston TX 77251, USA.
[317] MIDAS; University of California, San Francisco, CA 94143, USA.
[318] MACROMODEL; Columbia University, New York, NY 10027, USA.
[319] INSIGHT, DISCOVER; Biosym Technologies, San Diego, CA 92121, USA.
[320] COMPARE, PROXBUILDER, SPACFIL, CHEMLAB; Molecular Design Ltd., San Leandro, CA 94577, USA.
[321] AMBER; University of California, San Francisco, CA 94143, USA.
[322] HYDRA, CHARMM; Polygen Corp., Waltham, MA 02154, USA.
[323] ALCHEMY; Tripos Associates Inc., St. Louis MO 63117, USA.
[324] J. Cherfils, M.C. Vaney, I. Morize, E. Surcouf, N. Colloc'h, and J.P. Mornon, *J. Mol. Graphics*, **6**, 155 (1988).
[325] M.C. Marivet, J.-J. Bourguignon, C. Lugnier, A. Mann, J.-C. Stoclet, and C.-G. Wermuth, *J. Med. Chem.*, **32**, 1450 (1989).
[326] H. Shinkai, M. Nishikawa, Y. Sato, K. Toi, I. Kumashiro, Y. Seto, M. Fukuma, K. Dan, and S. Toyoshima, *J. Med. Chem.*, **32**, 1436 (1989).
[327] E. Pop, M.E. Brewster, J.J. Kaminski, and N. Bodor, *Int. J. Quantum Chem.*, **35**, 315 (1989).
[328] A.J. Hopfinger, *J. Amer. Chem. Soc.*, **102**, 7196 (1980).
[329] R. Langridge, T.E. Ferrin, I.D. Kuntz, and M.L. Connolly, *Science*, **211**, 661 (1981).
[330] N.L. Max, *IEEE Computer Graphics and Applications*, **3**, 21 (1983).
[331] P.A. Bash, N. Pattabiraman, C. Huang, T.E. Ferrin, and R. Langridge, *Science*, **222**, 1325 (1983).
[332] S. Namasivayam and P.M. Dean, *J. Mol. Graphics*, **4**, 46 (1986).
[333] C. Hansch and T.E. Klein, *Acc. Chem. Res.*, **19**, 392 (1986).
[334] P.M. Dean and P.-L. Chau, *J. Mol. Graphics*, **5**, 152 (1987).
[335] P.M. Dean and P. Callow, *J. Mol. Graphics*, **5**, 159 (1987).
[336] C. Thomson, D. Higgins, and C. Edge, *J. Mol. Graphics*, **6**, 171 (1988).
[337] V.A. Nicholson and G.M. Maggiora, *J. Math. Chem.*, **11**, 47 (1992).
[338] G.R. Marshall, C.D. Barry, H.E. Bosshard, R. Dammkoehler, and D.A. Dunn, "The Conformational Parameter in Drug Design: The Active Analog Approach", in *Computer Assisted Drug Design*, E.C. Olson and R.E. Christoffersen (Eds.), *ACS. Symp. Ser.*, **112**, 205 (1979).
[339] J.S. Richardson, *Adv. Protein Chem.*, **34**, 167 (1981).
[340] J.S. Richardson, *Methods in Enzymol.*, **115**, 341 (1985).
[341] R.E. Dickerson, in *The Proteins II,* H. Neurath (Ed.), Academic Press, New York, London, 1964, p.634

[342] P.D. Walker, G.A. Arteca, and P.G. Mezey, *J. Comput. Chem.*, **12**, 220 (1991).

[343] P.G. Mezey, *Theor. Chim. Acta*, **58**, 309 (1981).

[344] P.G. Mezey, *Theor. Chim. Acta*, **63**, 9 (1983).

[345] P.G. Mezey, "Reaction Topology and Quantum Chemical Molecular Design on Potential Energy Surfaces", in *New Theoretical Concepts for Understanding Organic Reactions*, J. Bertran and I.G. Csizmadia (Eds.), Nato ASI Series, Kluwer Academic Publishers, Dordrecht, 1989, pp 55-76.

[346] G.A. Arteca and P.G. Mezey, *J. Math. Chem.*, **10**, 329 (1992).

[347] F. Harary and P.G. Mezey, *J. Math. Chem.*, **2**, 377 (1988).

[348] G.A. Arteca and P.G. Mezey, *J. Math. Chem.*, **3**, 43 (1989).

[349] G.A. Arteca, G.A. Heal, and P.G. Mezey, *Theor. Chim. Acta*, **76**, 377 (1990).

[350] G.A. Arteca and P.G. Mezey, *Int. J. Quantum Chem.*, **38**, 713 (1990).

[351] P.G. Mezey, *J. Math. Chem.*, **8**, 91 (1991).

[352] G.S. Hammond, *J. Amer. Chem. Soc.*, **77**, 334 (1955).

[353] L. Melander, *The Transition State*, Royal Society Special Publications, Vol. 16, London, 1962.

[354] J. C. Polanyi, *J. Chem. Phys.*, **31**, 1338 (1959).

[355] M. H. Mok and J. C. Polanyi, *J. Chem. Phys.*, **51**, 1451 (1969).

[356] A. R. Miller, *J. Amer. Chem. Soc.*, **100**, 1984 (1978).

[357] N. Agmon, *J. Chem. Soc. Faraday Trans. II*, **74**, 388 (1978).

[358] J. R. Murdoch, *J. Amer. Chem. Soc.*, **105**, 2667 (1983).

[359] G.A. Arteca and P.G. Mezey, *J. Phys. Chem.*, **93**, 4746 (1989).

[360] G.A. Arteca and P.G. Mezey, *J. Comput. Chem.*, **9**,728 (1988).

[361] R.Carbó, L.Leyda, and M. Arnau, *Int. J. Quantum Chem.*, **17**, 1185 (1980).

[362] R. Carbó and Ll. Domingo, *Int. J. Quantum Chem.*, **32**, 517 (1987).

[363] R. Carbó and B. Calabuig, *Comput. Phys. Commun.*, **55**, 117 (1989).

[364] R. Carbó and B. Calabuig, *Int. J. Quantum Chem.*, **42**, 1681 (1992).

[365] R. Carbó and B. Calabuig, *Int. J. Quantum Chem.*, **42**, 1695 (1992).

[366] E.E. Hodgkin and W.G. Richards, *Int. J. Quantum Chem.*, **14**, 105 (1987).

[367] C. Burt, W.G. Richards, and P. Huxley, *J. Comput. Chem.*, **11**, 1139 (1990).

[368] M. A. Johnson, *J. Math. Chem.*, **3**, 117 (1989).

[369] M.A. Johnson and G.M. Maggiora (Eds.), *Concepts and Applications of Molecular Similarity*. Wiley, New York, 1990.

[370] P.G. Mezey (Ed.), *Mathematical Modeling in Chemistry*. VCH Publishers, New York, 1991.

[371] C. Soteros and S.G. Whittington, *J. Phys. A*, **21**, 2187 (1988).

[372] N. Madras, C. Soteros, and S.G. Whittington, *J. Phys. A*, **21**, 4617 (1988).

[373] S.G. Whittington, C. Soteros, and N. Madras, *J. Math. Chem.*, **7**, 87 (1991).

[374] S. Golomb, *Polyominoes*. Scribner's, New York (1965).

[375] F. Harary, "The cell growth problem and its attempted solutions", in *Beitrage zur Graphentheorie*. Teubner, Leipzig, 1968.

[376] M. Gardner, "Mathematical games in which players of ticktacktoe are taught to hunt for bigger game". *Scientific American* **240**, 18 (April 1979).

[377] G. Exoo and F. Harary, *Nat. Acad. Sci. Letters*, **10**, 67 (1987).

[378] F. Harary and M. Lewinter, *Int. J. of Comput. Math.*, **25**, 1 (1988).

[379] F. Harary and P.G. Mezey, *Theor. Chim. Acta*, **79**, 379 (1991).

[380] P.D. Walker and P.G. Mezey, *Int. J. Quantum Chem.*, **43**, 375 (1992).

[381] B. R. Gaines, *Int. J. Man-Mach. Stud.*, **8**, 623 (1976).

[382] L.A. Zadeh, *Inform. Control*, **8**, 338 (1965).

[383] L.A. Zadeh, *J. Math. Anal. Appl.*, **23**, 421 (1968).

[384] A. Kaufmann, *Introduction à la Théorie des Sous-Ensembles Flous.* Masson, Paris, 1973.

[385] L.A. Zadeh, *Theory of Fuzzy Sets,* in *Encyclopedia of Computer Science and Technology.* Marcel Dekker, New York, 1977.

[386] M.M. Gupta, R.K. Ragade, and R.R. Yager (Eds.), *Advances in Fuzzy Set Theory and Applications.* North-Holland, Leyden, 1979.

[387] D. Dubois and H. Prade, *Fuzzy Sets and Systems: Theory and Applications.* Academic Press, New York, 1980.

[388] E. Sanchez and M.M. Gupta (Eds.), *Fuzzy Information, Knowledge Representation and Decision Analysis.* Pergamon Press, London, 1983.

[389] E. Prugovecki, *Found. Phys.,* **4**, 9 (1974).

[390] E. Prugovecki, *Found. Phys.,* **5**, 557 (1975).

[391] E. Prugovecki, *J. Phys.* A, **9**, 1851 (1976).

[392] S.T. Ali and H.D. Doebner, *J. Math. Phys.,* **17**, 1105 (1976).

[393]   S.T. Ali and E. Prugovecki, *J. Math. Phys.,* **18**, 219 (1977).

[394] J. Maruani and P.G. Mezey, *J. Chim. Phys.,* **87**, 1025 (1990).

[395] J. Maruani and P.G. Mezey, *Int. J. Quantum Chem.,* **45**, 177 (1993).

[396] P.G. Mezey, *J. Math. Chem.,* **12**, 365 (1993).

[397] J.-E. Dubois, "Darc System in Chemistry", in *Computer Representation and Manipulation of Chemical Information.* W.T. Wipke, S. Heller, R. Fellmann, and E. Hyde (Eds.), Wiley, New York, 1974, p. 239.

[398] J.-E. Dubois, D. Laurent, and A. Aranda, *J. Chim. Phys.,* **70**, 1608 (1973).

[399] J.-E. Dubois, D. Laurent, and A. Aranda, *J. Chim. Phys.,* **70**, 1616 (1973).

[400] J.-E. Dubois, D. Laurent, P. Bost, S. Chambaud, and C. Mercier, *Eur. J. Med. Chem.,* **11**, 225 (1976).

[401] C. Mercier and J.-E. Dubois, *Eur. J. Med. Chem.,* **14**, 415 (1978).

[402] C. Mercier, Y. Sobel, and J.-E. Dubois, *Eur. J. Med. Chem.,* **16**, 473 (1981).

[403] J.-E. Dubois, C. Mercier, and Y. Sobel, *Compt. Rend. Acad. Sci. Paris,* **289C**, 89 (1979).

[404] J.-E. Dubois, *Proc. Int. Codata Conf.,* **8**, 155 (1982).

[405] J.-E. Dubois, J.-P. Doucet, and S.Y. Yue, in *Molecular Organization and Engineering.* J. Maruani (Ed.), Kluwer, Dordrecht, 1988, vol. 1, p. 173.

[406] C. Mercier, G. Trouiller, and J.-E. Dubois, *Quant. Struct.-Act. Rel.,* **9**, 88 (1990).

[407] P.D. Walker, PhD Thesis, University of Saskatchewan, 1992.

[408] G. A. Arteca and P.G. Mezey, *Biopolymers,* **32**, 1609 (1992).

[409] L.P. Hammett, *J. Amer. Chem. Soc.,* **59**, 96 (1937).

[410] L.P. Hammet, *Physical Organic Chemistry*, McGraw-Hill, New York, 1940 and 1970.

[411] P.W. Taft, in *Steric Effects in Organic Chemistry*. M.S. Newman (Ed.), Wiley, New York, 1956, p. 556.

[412] T. Fujita, J. Iwasa, and C. Hansch, *J. Amer. Chem. Soc.*, **36**, 5175 (1964).

[413] C. Hansch, *Acc. Chem. Res.*, **2**, 232 (1969).

[414] J.G. Topliss and P.J. Costello, *J. Med. Chem.*, **15**, 1066 (1972).

[415] H. Kubinyi, *J. Med. Chem.*, **20**, 625 (1977).

[416] J.C. Dearden and M.S. Townend, in *Herbicides and Fungicides - Factors Affecting their Activity*. N.R. McFarlane (Ed.), Chemical Society Special Publication No. 29, London, 1977, p. 135.

[417] G. Klopman, P. Andreozzi, and A.J. Hopfinger, *J. Amer. Chem. Soc.*, **100**, 6267 (1978).

[418] A. Warshel, *Proc. Natl. Acad. Sci. USA*, **75**, 5250 (1978).

[419] G.M. Crippen and T.F. Havel, *Acta Crystallogr. Sect. A*, **34**, 282 (1978).

[420] C. Hansch and A.J. Leo, *Substituent Constants for Correlation Analysis in Chemistry and Biology*. Wiley, New York, 1979.

[421] H. Kubinyi, in *Progress in Drug Research*, Vol. 23. E. Jucker (Ed.), Birkhäuser, Basel, 1979, p. 97.

[422] J.C. Dearden and M.S. Townend, *Pestic. Sci.*, **10**, 87 (1979).

[423] J.T. Chou and P.C. Jurs, *J. Chem. Inf. Comput. Sci.*, **19**, 172 (1979).

[424] I. Lukovits and A. Lopata, *J. Med. Chem.*, **23**, 443 (1980).

[425] G.M. Crippen, *J. Med. Chem.*, **23**, 599 (1980).

[426] P. Li, C. Hansch, D. Matthews, J.M. Blaney, P. Langridge, T.J. Delcamp, S.S. Susten, and J.H. Freisheim, *Quant. Struct.-Act. Relat.*, **1**, 1 (1982).

[427] G.W.A. Milne, C.L. Fiske, S.R. Heller, and R. Potenzone, *Science*, **215**, 371 (1982).

[428] I. Lukovits, *J. Med. Chem.*, **26**, 1104 (1983).

[429] A. Burger, *A Guide to the Chemical Basis of Drug Design*, Wiley, New York, 1983.

[430] T. Fujita, *Progr. Phys. Org. Chem.*, **14**, 75 (1983).

[431] P.P. Mager, *Multidimensional Pharmacochemistry, Design of Safer Drugs*. Academic Press, New York, 1984.

[432] S. Ungar, in *QSAR in Design of Bioactive Molecules*. M. Kuchar (Ed.), J.R. Prous Publishers, Barcelona, 1984.

[433] J.K. Seydel (Ed.), *QSAR and Strategies in the Design of Bioactive Compounds*. VCH Publishers, Weinheim, Germany, 1985.

[434] J. Gasteiger, M.G. Hutchings, M. Marsili, and H. Saller, "Rapid Calculation of Electronic Effects in Organic Molecules", in *QSAR and Strategies in the Design of Bioactive Compounds*. J.K. Seydel (Ed.) VCH Publishers, Weinheim, Germany, 1985, pp. 80-97.

[435] F. Darvas, A. Lopata, Z. Budai, and L. Petöcz, "Computer Assisted Design of a Novel Type of Tranquillant" in *QSAR and Strategies in the Design of Bioactive Compounds*. J.K. Seydel (Ed.), VCH Publishers, Weinheim, Germany, 1985, pp. 324-327.

[436] A. Lopata, F. Darvas, A. Stadler-Szöke, and J. Szejtli, "Quantitative Structure-Stability Relationships Among Inclusion Complexes of Cyclodextrins. Part 2. Steroid Hormones", in *QSAR and Strategies in the*

Design of Bioactive Compounds. J.K. Seydel (Ed.), VCH Publishers, Weinheim, Germany, 1985, pp. 353-356.

[437] B. Bordás, "Stepwise Canonical Correlation Analysis: A New Approach for the Study of Structural Requirements of Broad Spectrum Activity", in *QSAR and Strategies in the Design of Bioactive Compounds*. J.K. Seydel (Ed.), VCH Publishers, Weinheim, Germany, 1985, pp. 389-392.

[438] Z. Dinya, T. Timár, S. Hosztafi, A. Fodor, P. Deák, A. Somogyi, and M. Berényi, "Theoretical Considerations About the Possible Mode of Action of Precocenes", in *QSAR and Strategies in the Design of Bioactive Compounds*. J.K. Seydel (Ed.), VCH Publishers, Weinheim, Germany, 1985, pp. 403-409.

[439] L. Meyler and A. Hexheimer, *Side Effects of Drugs,* Vol. VII. Excerpta Medica, Amsterdam, 1972.

[440] E.J. Lien and L.L. Lien, *J. Clin. Hosp. Pharm.*, **5**, 255 (1980).

[441] E.J. Lien and R.T. Koda, *Drug. Intell. Clin. Pharm.*, **15**, 434 (1981).

[442] E.J. Lien, *SAR: Side Effects and Drug Design*. Marcel Dekker, New York, 1987.

[443] C.D. Zachmann, W. Heiden, M. Schlenkrich, and J. Brickmann, *J. Comp. Chem.*, **13**, 76 (1992).

[444] P.G. Mezey, *J. Math. Chem.*, **4**, 377 (1990).

[445] X. Luo, G. A. Arteca, and P.G. Mezey, *Int. J. Quantum Chem.*, **42**, 459 (1992).

[446] H.J. Bandelt and A.W.M. Dress, *Bull. Math. Biology*, **51**, 133 (1989).

[447] H. Bock, K. Ruppert, C. Näther, Z. Havlas, H.-F. Herrmann, C. Arad, I. Göbel, A. John, J. Meuret, S. Nick, A. Rauschenbach, W. Seitz, T. Vaupel, and B. Solouki, *Angew. Chem. Int. Ed. Engl.*, **31**, 550 (1992).

[448] H. Zabrodsky, S. Peleg, and D. Avnir, *J. Amer. Chem. Soc.*, **114**, 7843 (1992).

[449] R. Cammi and E. Cavalli, *Acta Cryst.*, **B48**, 245 (1992).

[450] M. Senechal, *Crystalline Symmetries, An Informal Introduction*. Adam Hilger Publishers, Bristol, U.K., 1990, pp. 70.

[451] S. Iwata (Material Science Editor), *Forma, Continuation of Science on Form*, University of Tokyo, personal communication.

[452] P.G. Mezey, *J. Amer. Chem. Soc.*, **112**, 3791 (1990).

[453] P.G. Mezey, *J. Math. Chem.*, **3**, 407 (1989).

[454] D. Hilbert and S. Cohn-Vossen, *Anschauliche Geometrie,* Springer, Berlin, 1932.

INDEX